Muscle Fatigue Mechanisms in Exercise and Training

Medicine and Sport Science

Vol. 34

Series Editors
M. Hebbelinck, Brussels
R.J. Shephard, Toronto, Ont.

Founder and Editor from 1969 to 1984
E. Jokl, Lexington, Ky.

KARGER

Basel · München · Paris · London · New York · New Delhi · Bangkok · Singapore · Tokyo · Sydney

Proceedings of the 4th International Symposium on Exercise and Sport Biology,
Nice, November 1–4, 1990

Muscle Fatigue Mechanisms in Exercise and Training

Volume Editors

P. Marconnet, Nice
P.V. Komi, Jyväskylä
B. Saltin, Copenhagen
O.M. Sejersted, Oslo

92 figures and 19 tables, 1992

KARGER

Basel · München · Paris · London · NewYork · New Delhi · Bangkok · Singapore · Tokyo · Sydney

Medicine and Sport Science

Published on behalf of the
International Council of Sport Science and Physical Education

Library of Congress Cataloging-in-Publication Data
International Symposium on Exercise and Sport Biology (4th: 1990: Nice, France)
Muscle fatigue mechanisms in exercise and training: proceedings of the 4th International
Symposium on Exercise and Sport Biology, Nice, November 1–4, 1990/volume editor,
P. Marconnet.
(Medicine and sport science; vol. 34)
Includes bibliographical references and index.
1. Muscles — Physiology — Congresses. 2. Fatigue — Congresses.
3. Exercise — Physiological aspects — Congresses. I. Marconnet, P. (Pierre)
II. Title. III. Series.
[DNLM: 1. Exercise — physiology — congresses. 2. Fatigue — congresses.
3. Muscles — physiology — congresses. 4. Sports Medicine — congresses.
ISBN 3–8055–5483–4(alk. paper)

Contents

Preface

The 4th Nice Symposium on Biology of Exercise and Physical Training, the proceedings of which are included in this volume, aimed, as did former symposia at: (i) providing the most recent and advanced information on a specific subject – here *Muscle Fatigue* – by those researchers who are amongst the best in their field; (ii) favoring contact and exchange between scientists, as well as the latter and users in the field; (iii) creating a scientific event which is able to attract the interest of various groups of people such as scientists and technicians, athletes, medical doctors, reporters, civil servants, and politicians, to the importance of *Exercise Biology,* an area which at the moment is still underdeveloped in France.

During his opening presentation, Bengt Saltin took the opportunity to remind us that *Muscle Fatigue* is currently still a subject of considerable interest, although it has been addressed by researchers for more than a century.

The scientific committee paid attention to organizing the content of the Symposium in such a way as to concede a large part to the more recent fundamental data basically involved in the mechanisms of muscle fatigue without minimizing the methodological aspects, addressed through presentations of various experimental models ranging from isolated muscle fiber to the whole body. It is of interest to note that alternative hypotheses to the more classical theories about local fatigue received particular mention in the program.

It is wished that, while reading this volume, all the participants of this Symposium will remember the fascinating oral presentations and discussions. This book is also devoted to all those who were unable to participate, but are interested in physical activity and muscle fatigue, and who will find the main substance in each chapter.

The organizers are indebted for the encouragement and financial support of the following institutions: Ministère et Direction Régionale de la Jeunesse et des Sports français, Municipalité de Nice, Université de Nice-Sophia-Antipolis, Conseil Général des Alpes Maritimes, Conseil Régional de la Région PACA. The staff whose work contributed strongly to the success of the meeting deserves special mention.

Pierre Marconnet
Nice, November 1991

Marconnet P, Komi PV, Saltin B, Sejersted OM (eds): Muscle Fatigue Mechanisms in Exercise and Training. Med Sport Sci. Basel, Karger, 1992, vol 34, pp 1–10

ATP-Dependent K+ Channels and Other K+ Channels of Muscle: How Exercise May Modulate Their Activity

Noel Wyn Davies

Ion Channel Group, Department of Physiology, University of Leicester, UK

Introduction

There have been extensive studies on the effect that vigorous muscular activity has on the build up of metabolites and on the resulting contractile performance of muscle fibers. Many factors change during muscle fatigue ranging from a loss of force-generating capacity to biochemical changes such as a decline in the concentration of creatine phosphate and adenosine triphosphate (ATP) and an increase in the concentration of inorganic phosphate and protons [1]. During exercise there is an accumulation of extracellular K+, which seems to occur solely as a consequence of increased K+ efflux from skeletal muscle during exercise [2]. During intense activity the conductance properties of individual muscle fibers change, with the most noticeable effect being an increase in the resting K+ conductance first seen in exhausted frog skeletal muscle fibers by Fink and Lüttgau [3]. In this chapter I will concentrate on this observation and propose a mechanism whereby the increased K+ conductance is linked to the fall in intracellular pH (pH_i) that occurs during severe exercise.

K+ Channels Present in Skeletal Muscle

Three main types of K+ channels exist in skeletal muscle, the delayed rectifier, the inward rectifier and ATP-dependent K+ channels (K_{ATP} channels). Some of the basic properties of these three types of K+ channels are listed in table 1. It is noteworthy that, despite quite different activation and

Table 1. Some basic properties of K^+ channels found in skeletal muscle

Channel type	Properties and pharmacology
Delayed rectified K^+ channels	Activated by depolarization but maintained depolarization leads to complete inactivation Blocked by TEA^+, Cs^+, Ba^{2+}, H^+, 4-AP, quinidine
Inward rectifier K^+ channels	Activated by hyperpolarization and their gating depends on external $[K^+]$ Blocked by TEA^+, Cs^+, Rb^+, Na^+, Ba^{2+}, and Sr^{2+}
ATP-dependent K^+ channels	Relatively voltage insensitive Internal ATP inhibits their activity Blocked by TEA^+, Cs^+, Ba^{2+}, 4-AP, and glibenclamide*

The blocking agents listed are those which can act from the outside.
TEA^+ = Tetraethylammonium ions; 4-AP = 4-aminopyridine. Data obtained from Stanfield [27], Hille [28], and *Beeston, Davies & Stanfield [unpubl. observations].

inactivation properties, these K^+ channels share similar pharmacological properties. To date Ca^{2+}-activated K^+ channels have not been seen in adult skeletal muscle [4], although Latorre et al. [5] have extracted a K^+ channel from rabbit muscle which shows Ca^{2+} dependence when incorporated into lipid bilayers.

The function of delayed rectifier channels, as in neurones, is to repolarize the muscle membrane during an action potential. The fast activation kinetics of the delayed rectifier ensures that action potentials of vertebrate fast skeletal muscle are kept short. The role of the inward rectifier in skeletal muscle is unclear although possibilities include a method to limit K^+ accumulation in the T-tubular system by favoring K^+ entry rather than exit, particularly if the membrane potential is driven more negative than the equilibrium potential of K^+ (E_K) in the T-tubules by virtue of the conductances of the surface membrane.

Using the patch clamp technique [6], Noma [7] first demonstrated the existence of K^+ channels in heart cells that were sensitive to the intracellular concentration of ATP. These K_{ATP} channels have subsequently been seen in pancreatic β-cells [8], amphibian and mammalian skeletal muscle [9, 10], neurones [11] and smooth muscle cells [12]. K_{ATP} channels are present in high densities in the membrane of skeletal muscle, roughly equal to the density of delayed rectifier K⁺ channels [9]. In contrast to delayed rectifier channels, they do not have a strong voltage-dependence, and since they do not inactivate they can be active at the resting membrane potential given the right conditions. Indeed, evidence suggests that the increased K⁺ conductance of exhausted muscle fibers is due to the activation of these channels (see below). However, their open probability (P_{open}) at rest is likely to be very low.

The Effect of Exercise on Skeletal Muscle K⁺ Channels

Studies on the effect of exercise on the membrane properties of skeletal muscle have been made by exhausting the muscle by repetitive stimulation and in some cases this exhaustion procedure has been preceded by metabolically poisoning the muscle. Fink and Lüttgau [3] demonstrated that the K⁺ conductance of exhausted frog skeletal muscle fibers increased 130-fold and that there was a slight increase in the Cl⁻ conductance, although it could be argued that increases of this magnitude may never be reached in vivo. With progressive exhaustion, the action potential amplitude decreased until eventually no propagation of the action potential was possible. The explanation of these results centered on the types of K⁺ conductances known at that time, either suggesting a modification of voltage-dependent K⁺ conductances or the activation of Ca²⁺-activated K⁺ channels [13].

Castle and Haylett [14] re-examined the K⁺ conductance of exhausted muscle fibers to determine whether the channels involved could be directly identified with any of those already described for skeletal muscle, including K_{ATP} channels. By investigating the actions of known K⁺ channel blockers they concluded that a substantial part of the K⁺ conductance activated in metabolically exhausted muscle is due to the activation of K_{ATP} channels. It seems likely, therefore, that K_{ATP} channels are activated by the changing metabolic state of the muscle cell caused by severe exercise.

Table 2 shows the major changes occurring in cell metabolites with exercise. It was originally proposed that the fall in ATP concentration due to metabolic exhaustion would lead to the activation of these channels [14]. However, as table 2 shows, relatively little change occurs in the concentration of ATP during exercise [15], and certainly not enough to explain the marked increase in K⁺ conductance observed. One of the most marked effects of exhaustive exercise is a fall in pH_i. A decrease of almost 1 pH unit has been recorded with pH-sensitive electrodes in frog skeletal muscle [16] and in

Table 2. Metabolic changes occurring during vigorous exercise

Type of exercise	Changes in metabolites
Sustained contraction	pH_i less than 6.6 75–90% decline in CrP 20–30% decline in ATP 20–50% increase in ADP 300% increase in P_i
Exhaustive dynamic exercise	pH_i less than 6.6 65% decline in CrP 30% decline in ATP external $K^+ = 8$ mM

CrP = Creatine phosphate; P_i = inorganic phosphate. Data obtained from Latorre et al. [5].

human muscle using NMR spectroscopy [17]. This prompted an investigation of the effect of changing pH_i on the activity of K_{ATP} channels and on their sensitivity to ATP.

The Effect of pH$_i$ on K$_{ATP}$ Channels

K_{ATP} channels have complicated kinetics, even in the absence of ATP. In frog skeletal muscle they have at least two open and four closed states, and openings occur in bursts. In the absence of ATP the kinetics of K_{ATP} channels are altered by changing pH_i despite the fact that P_{open} remains virtually constant [18, 19]. Another effect of decreasing pH_i is a reduction in the single channel current. The main kinetic changes that occur on lowering pH_i are an increase in the number of brief closed events, a slight increase in mean open time and an increase in burst duration.

At pH_i of 7.2, ATP inhibits the activity of K_{ATP} channels with a K_i of 17 μM [Davies, Standen & Stanfield, unpubl.], which is much less than the concentration of ATP present in muscle. This inhibition is not associated with the hydrolysis of ATP [4]. A decrease in pH_i to 6.3 increases this K_i value 13-fold to over 200 μM. Thus, lowering pH_i in the presence of ATP removes much of the inhibiting power of ATP and effectively activates these channels. This effect also occurs in the presence of excess Mg^{2+} ions [19], although absolute values of K_i in the presence of Mg^{2+} have not been measured in skeletal muscle. Figure 1 shows the effect of pH_i 7.2 and 6.3 on

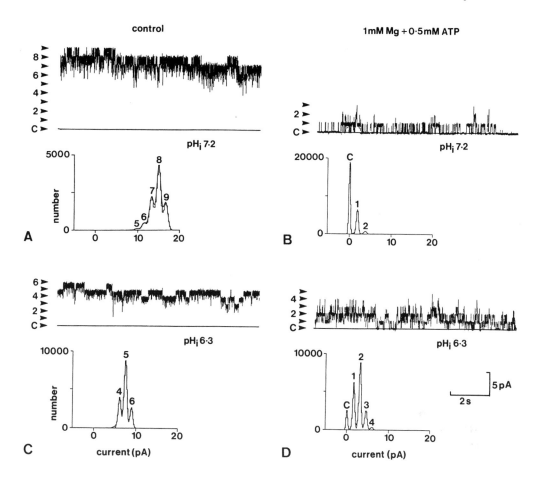

Fig. 1. The activity of K$_{ATP}$ channels in the absence and presence of 1 mM Mg^{2+} and 0.5 mM ATP at pH$_i$ 7.2 *(A, B)* and 6.3 *(C, D)*. The records were obtained from an isolated inside-out patch of frog skeletal muscle. The arrows at the side of each trace indicate the level at which 0 *(C)*, 1, 2, 3, ... channels are open. The inhibitory effect of Mg^{2+} and ATP was much greater at pH$_i$ 7.2 than at 6.3. The activity in *B* normalized to *A* was 0.05 while that in *D* normalized to *C* was 0.20. Below each record is an amplitude histogram of a 7s under each condition. The peaks, which correspond to the number of channels open, were fitted with a Gaussian distribution. These fits give single-channel amplitudes of 1.90, 1.89, 1.48 and 1.50 pA in *A, B, C* and *D* respectively. Run-down of channel activity occurred between parts *A* and *D* which was not associated with the effect of pH$_i$. The holding potential was −3 mV, [K$^+$]$_o$ = 10 mM and [K$^+$]$_i$ = 120 mM. Reproduced with permission from Davies [19].

K_{ATP} channel activity in the absence and presence of Mg^{2+} and ATP. These results were recorded from an inside-out patch of sarcolemmal membrane from frog skeletal muscle. Note that the number of active channels had decreased from 9 to 6 in the interval between parts A and C, this effect was due to 'run-down' of the channels and is not associated with the effect of pH. It is evident that the block of the channels by Mg^{2+} and ATP is much less effective at the lower pH value. Mean fractional block of K_{ATP} channels by 1 mM Mg^{2+} and 0.5 mM ATP at different pH_i values are shown in figure 2.

What Causes the Effect of pH_i on K_{ATP} Channels?

Several possibilities exist as to the cause of the pH_i actions described above including the following three cases: (i) the properties of the ATP-channel complex is itself altered by changing pH_i; (ii) only certain ATP complexes bind to the site and that decreasing pH_i produces a sufficient reduction in the concentration of the active complexes, and (iii) the binding site for ATP is altered by changing pH_i. Case (i) is unlikely since at all pH_i values tested, saturating concentrations of ATP could completely inhibit the activity of K_{ATP} channels. This means that when ATP is bound to the channel it is unable to open. The second case is quite likely to occur to an extent although the changes in the concentrations of the various ATP complexes are not sufficient to account for the change in blocking efficiency. Furthermore, the persistence of the effect in the presence of excess Mg^{2+}, in which alterations of pH have a much smaller effect on the ATP complexes, also indicates that the sensitivity of the ATP complexes to pH is not the sole cause of the actions of pH_i. The most plausible explanation is a combination of the latter two cases in which the binding site for ATP is altered by changing pH_i and that the concentration of the most effective ATP complex is reduced by decreasing pH_i.

Comparison of the Effect of pH_i on K_{ATP} Channels of Skeletal and Cardiac Muscle

Similar experiments have been performed on K_{ATP} channels in isolated inside-out patches of cardiac cells by Lederer and Nichols [20]. In contrast to the results described here for skeletal muscle, decreasing pH_i under similar conditions has a much less marked effect on cardiac K_{ATP} channels. The stoichiometry of the block of frog skeletal muscle K_{ATP} channels by ATP is

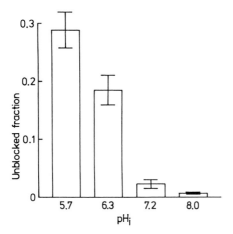

Fig. 2. The blocking efficiency of 1 mM Mg^{2+} and 0.5 mM ATP on K$_{ATP}$ channels of frog skeletal muscle at different pH$_i$ values. The bars show the mean remaining unblocked current (normalized) ±SEM. The data are pooled from 16 patches where the holding potential was –3 mV, [K⁺]$_o$ = 10 mM and [K⁺]$_i$ = 120 mM.

consistent with 1:1 binding at all pH$_i$ values tested. In contrast to this, Lederer and Nichols [20] found that the blocking stoichiometry in heart was 2:1 at pH$_i$ 7.25 with a K$_i$ of 25 μM but 3:1 with a K$_i$ of 50 μM at pH$_i$ 6.25.

The Effect of pH$_i$ on Other K⁺ Channels

K$_{ATP}$ channels are unusual in that a decrease in pH$_i$ effectively increases their activity. The activity of most other K⁺ channels in a number of preparations is reduced by decreasing pH$_i$. These include the inward rectifier of frog skeletal muscle and the delayed rectifier of crayfish muscle [21]. Although the details of these effects have not yet been investigated, the effect of decreasing pH$_i$ on the overall K⁺ conductance of skeletal muscle is likely to be a linearization of the current-voltage relation which has already been observed in exhausted frog muscle [13].

The Effect of Other Metabolites on K$_{ATP}$ Channels

Other metabolites which also vary with exercise include a rise in both ADP and inorganic phosphate concentration. In frog skeletal muscle ADP inhibits channel activity but with a much higher K_i (0.44 mM) than ATP [4]. In heart muscle in the presence of Mg^{2+}, ADP opposes the inhibition induced by ATP [20, 22, 23]. In skeletal muscle, however, no evidence for this regulation has been found either in the presence or absence of Mg^{2+} [4; Davies, Standen & Stanfield, unpubl. observations]. This contrast is likely to be associated with the difference in the stoichiometry of ATP binding to cardiac and skeletal muscle K_{ATP} channels [20, 24]. The inhibitory efficiency of 0.2 mM ATP was also found to be unaffected by the presence of 2 mM NaH_2PO_4 [4]; however, a higher concentration of inorganic phosphate has not been tested.

One factor which may have more subtle effects is the accumulation of external K^+. It has already been mentioned that the gating kinetics of the inward rectifier channels is sensitive to external $[K^+]$. This is also true for K_{ATP} channels in both skeletal and cardiac muscle [25, 26]. Thus, the increase in extracellular K^+ which occurs during exercise [2] may have a modulatory effect on K_{ATP} and other K^+ channels of skeletal muscle.

Conclusions

In skeletal muscle the increase in K^+ conductance which occurs with exhaustion appears to be associated mainly with an increase in the activity of K_{ATP} channels. A likely candidate linking the activity of these channels with the metabolic state of the muscle cell is pH_i, although other additional factors cannot be excluded. A fall in pH_i of about 1 unit (a 10-fold increase in $[H^+]$), which is known to occur during severe exercise, will lead to an approximately 10-fold increase in the P_{open} of K_{ATP} channels. Since K_{ATP} channels are present at high densities in skeletal muscle membrane it is possible that a substantial K^+ current flows through these channels during the latter part of vigorous exercise. Opening of these channels may reduce the excitability of muscle fibers and cause them to rest. However, further experiments are necessary to clarify the role of K_{ATP} channels in muscle fatigue.

References

1   Vøllestad NK, Sejersted OM: Biochemical correlates of fatigue. Eur J Appl Physiol 1988;57:336–347.
2   Medbø JI, Sejersted OM: Plasma potassium changes with high intensity exercise. J Physiol 1990;421:105–122.

3 Fink R, Lüttgau HCh: An evaluation of the membrane constants and the potassium conductance in metabolically exhausted muscle fibres. J Physiol 1976;263:215–238.

4 Spruce AE, Standen NB, Stanfield PR: Studies of the unitary properties of adenosine-5′-triphosphate-regulated potassium channels of frog skeletal muscle. J Physiol 1987;382:213–236.

5 Latorre R, Vergara C, Hidalgo C: Reconstitution in planar lipid bilayers of Ca^{2+}-dependent K$^+$ channels from rabbit skeletal muscle. Proc Natl Acad Sci USA 1982; 79:805–809.

6 Hamill OP, Marty A, Neher E, Sakmann B, Sigworth SJ: Improved patch-clamp techniques for high-resolution current recording from cells and cell-free membrane patches. Pflügers Arch 1981;391:85–100.

7 Noma A: ATP-regulated K$^+$ channels in cardiac muscle. Nature 1983;305:147–148.

8 Cook DL, Hales CN: Intracellular ATP directly blocks K$^+$ channels in pancreatic B-cells. Nature 1984;311:271–273.

9 Spruce AE, Standen NB, Stanfield PR: Voltage-dependent, ATP-sensitive potassium channels of skeletal muscle membrane. Nature 1985;316:736–738.

10 Burton F, Dörstelmann U, Hutter OF: Single channel activity in sarcolemmal vesicles from human and other mammalian muscles. Muscle Nerve 1988;11:1029–1038.

11 Ashford MLJ, Sturgess NC, Trout NJ, Gardner NJ, Hales CN: Adenosine-5′-triphosphate-sensitive ion channels in neonatal rat cultured central neurones. Pflügers Arch 1988;412:297–304.

12 Standen NB, Quayle JM, Davies NW, Brayden JE, Huang Y, Nelson MT: Hyperpolarizing vasodilators activate ATP-sensitive K$^+$ channels in arterial smooth muscle. Science 1989;245:177–180.

13 Fink R, Hase S, Lüttgau HCh, Wettwer E: The effect of cellular energy reserves and internal calcium ions on the potassium conductance in skeletal muscle of the frog. J Physiol 1983;336:211–228.

14 Castle NA, Haylett DG: Effect of channel blockers on potassium efflux from metabolically exhausted frog skeletal muscle. J Physiol 1987;383:31–43.

15 Carlson FD, Siger A: The mechanochemistry of muscular contraction. I. The isometric twitch. J Gen Physiol 1960;44:33–60.

16 Renaud JM: The effect of lactate on intracellular pH and force recovery of fatigued sartorius muscles of the frog Rana pipiens. J Physiol 1989;416:31–47.

17 Pan JW, Hamm JR, Rothman DL, Shulman RG: Intracellular pH in human skeletal muscle by ^{1}H NMR. Proc Natl Acad Sci USA 1988;85:7836–7839.

18 Davies NW, Standen NB, Stanfield PR: The kinetic behaviour of K$_{ATP}$ channels in isolated membrane patches of frog skeletal muscle is regulated by internal pH. J Physiol 1989;418:188P.

19 Davies NW: Modulation of ATP-sensitive K$^+$ channels in skeletal muscle by intracellular protons. Nature 1990;343:375–377.

20 Lederer WJ, Nichols CG: Nucleotide modulation of the activity of rat heart ATP-sensitive K$^+$ channels in isolated membrane patches. J Physiol 1989;419:193–211.

21 Moody M Jr: Effects of intracellular H$^+$ on the electrical properties of excitable cells. Ann Rev Neurosci 1984;7:257–278.

22 Findlay I: ATP^{4-} and ATP-Mg inhibit the ATP-sensitive K$^+$ channel of rat ventricular myocytes. Pflügers Arch 1988;412:37–41.

23 Findlay I: Effects of ADP upon the ATP-sensitive K$^+$ channel in rat ventricular myocytes. J Membr Biol 1988;101:83–92.

24 Davies NW, Standen NB, Stanfield PR: ATP-dependent potassium channels of muscle cells: their properties, regulation, and possible functions. J Biomembr Bioenerg 1991;23:509–535.

25 Quayle JM: Kinetics and block of ATP-sensitive K^+ channels of frog skeletal muscle; thesis, University of Leicester, 1988.

26 Zilberter Y, Burnashev N, Papin A, Portnov V, Khodorov B: Gating kinetics of ATP-sensitive single potassium channels in myocardial cells depends on electromotive force. Pflügers Arch 1988;411:584–589.

27 Stanfield PR: Tetraethylammonium ions and the potassium permeability of excitable cells. Rev Physiol Biochem Pharmacol 1983;97:1–67.

28 Hille B: Ionic Channels of Excitable Membranes. Sunderland, Sinauer, 1984.

Noel Wyn Davies, Ion Channel Group, Department of Physiology,
University of Leicester, PO Box 138, GB–Leicester LE1 9HN (UK)

Marconnet P, Komi PV, Saltin B, Sejersted OM (eds): Muscle Fatigue Mechanisms in Exercise and Training. Med Sport Sci. Basel, Karger, 1992, vol 34, pp 11–19

Regulation of Ca²⁺ Release from Skeletal Muscle Sarcoplasmic Reticulum[1]

Annegret Herrmann-Frank[a], Gerhard Meissner[b]

[a]Department of Cell Physiology, Ruhr-Universität Bochum, FRG; [b]Department of Biochemistry and Biophysics, University of North Carolina, Chapel Hill, N.C., USA

Introduction

The release of Ca^{2+} ions from the intracellular membrane compartment sarcoplasmic reticulum (SR) is an essential step in the activation of striated and smooth muscle. In skeletal muscle, Ca^{2+} release is initiated by a surface membrane action potential which is propagated by the transverse tubular system (T system) into the cell interior. By a complex process not yet fully understood, the excitatory electrical signal triggers the rapid release of Ca^{2+} ions from sarcoplasmic reticulum. Binding of the released Ca^{2+} to regulatory proteins associated with the contractile filaments leads to contraction of muscle fiber. Relaxation occurs when the released Ca^{2+} ions are again sequestered by the SR via the action of a Ca^{2+} pump or Ca^{2+}/Mg^{2+}-ATPase.

Signal transmission from the T system to SR takes place where the two membranes come into close contact and a narrow gap of 12–18 nm is traversed by irregular periodic densities variously named 'feet' [4], 'bridging structures' [18], or 'spanning proteins' [1]. On the molecular level, two proteins have been shown to be involved in signal transmission: the dihydropyridine (DHP) receptor in the T tubular membrane and the ryanodine receptor in the SR membrane. Biophysical and biochemical evidence has indicated that the DHP receptor functions as the voltage sensor of 'excitation-contraction (EC) coupling' in skeletal muscle [14] while the ryanodine

[1] Supported by NIH and MDA grants and DFG fellowship (A.H.F.)

Table 1. Properties of skeletal muscle Ca^{2+} release channel

1	Composition	30S homotetramer of M_r -2.3 × 10^6 (four -565,000 subunits)
2	Structure	Four-leaf clover (quatrefoil)
3	Conductance	100 pS with 50 mM Ca^{2+} 600 pS with 500 mM Na^+
4	Regulation	Activation by μM Ca^{2+}, and modulation by mM ATP, mM Mg^{2+}, μM calmodulin and pH
5	Exogenous ligands	Activation by nM ryanodine and mM caffeine Inhibition by μM ryanodine

receptor has been identified as the Ca^{2+} release channel of sarcoplasmic reticulum [3, 8]. According to the mechanical coupling hypothesis [16], an action potential induces a conformational change in the skeletal muscle DHP receptor protein. This protein conformational change is then directly sensed by the SR feet structures causing the opening of the Ca^{2+} release channel.

In this article, we summarize studies we have performed to better understand the mechanism of SR Ca^{2+} release in mammalian skeletal muscle. We have found that rapid SR Ca^{2+} release is mediated by a 30S Ca^{2+} release channel complex which is activated by Ca^{2+} and modulated by ATP, Mg^{2+}, calmodulin, pH, and the two Ca^{2+}-releasing drugs caffeine and ryanodine.

Molecular Properties of the Skeletal Muscle Ca^{2+} Release Channel Complex

Identification and subsequent purification and structural definition of the SR Ca^{2+} release channel in skeletal muscle has been greatly helped by the availability of [^{3}H]ryanodine as a channel specific marker [8]. Table 1 summarizes presently known properties of the rabbit skeletal SR Ca^{2+} release channel as determined in our laboratory. The Chaps-solubilized Ca^{2+} release channel has been isolated by density gradient centrifugation as a 30S protein complex comprised of four identical polypeptides of apparent relative molecular mass of $\sim$400,000 on sodium dodecyl sulfate (SDS) polyacrylamide gels. Cloning and sequencing of the complementary DNA of the rabbit and human muscle ryanodine receptors has indicated an open reading frame

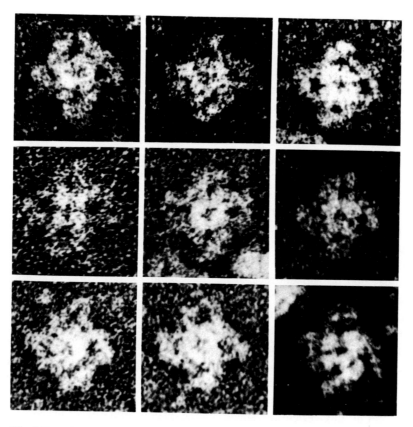

Fig. 1. Negative-stain electron micrograph of the 30S Ca²⁺ release channel complex. Dimensions of the four-leaf clover-like structures are 34 nm from the tip of one leaf to the tip of the opposite leaf. The central electron dense region has a diameter of ~14 nm with a central 'hole' of 1–2 nm diameter. Taken with permission from Lai et al. [9].

of 15.2 kb encoding 5,035 amino acids with a relative molecular mass ~565,000 [19, 20]. Electron microscopy of negatively stained samples has revealed a four-leaf clover-like (quatrefoil) arrangement of the four 565,000 polypeptides (fig. 1). A major portion of the Ca²⁺ release channel complex has been predicted to be cytoplasmically located and to be identical with the protein bridges that span the narrow gap between T system and SR membranes. ⁴⁵Ca²⁺ flux vesicle and single channel-planar lipid bilayer measurements have indicated that the main regulatory sites are situated in the cytoplasmic portion of the channel.

Regulation of Ca²⁺ Release Channel Activity

Regulation of SR Ca²⁺ release channel activity has been studied in our laboratory (1) by determining the $^{45}Ca^{2+}$ efflux behavior of 'heavy' SR vesicles, and (2) by incorporating the purified 30S Ca²⁺ release channel complex into planar lipid bilayers. The single channel recording of figure 2 depicts several of the fundamental functional properties of the SR Ca²⁺ release channel. The detergent-solubilized 30S protein complex was purified on a sucrose gradient and then directly incorporated into a planar lipid bilayer formed by painting a solution of phospholipid in n-decane across a 0.3-μm diameter aperture. Channel activities and ion currents were monitored using a voltage clamp circuit [17]. The bilayer separated two aqueous buffers, both of which contained 500 mM Na⁺ as the conducting ion and 6 μM free Ca²⁺ to partially activate the Ca²⁺ release channel. Under these conditions, we observed a rapidly gating channel with a large conductance of about 600 pS (upper trace of fig. 2). Channel activity was decreased by lowering the free Ca²⁺ concentration from 6 μM to 20 nM in the cis chamber (SR cytoplasmic side), and again increased by the addition of 2 mM ATP cis. Perfusion of the trans chamber (SR luminal side) with a 50 mM Ca²⁺ buffer and the cis chamber with the impermeable cation Tris⁺ resulted in a Ca²⁺ conductance of ∼90 pS and reversal potential of +30 mV. Under the same buffer conditions, both conductance and reversal potential are typical for the native Ca²⁺ release channel [17].

$^{45}Ca^{2+}$ flux-vesicle and single channel-planar lipid bilayer measurements have indicated that the isolated Ca²⁺ release channel is activated by Ca²⁺ and ATP, and inhibited by Mg²⁺ and calmodulin. The $^{45}Ca^{2+}$ efflux behavior of rabbit skeletal SR vesicles has been studied in the presence of Ca²⁺, Mg²⁺, adenine nucleotide and calmodulin in the release medium, using a rapid mixing-quench protocol [12]. Table 2 shows that at $10^{-8} M$ Ca²⁺, $^{45}Ca^{2+}$ release was slow, occurring with a first order rate constant of 0.1 s⁻¹. Increase in free Ca²⁺ to $2 \times 10^{-6} M$ and addition of the nonhydrolyzable ATP analog AMP-PCP to the release medium increased the release rate by a factor of nearly 1000. Intermediate efflux rates were observed when the Ca²⁺ release channel was activated at $4 \times 10^{-6} M$ free Ca²⁺ or by the addition of 5 mM ATP to the $10^{-8} M$ Ca²⁺ release medium. Mg²⁺ had an inhibitory effect when added to nanomolar and micromolar Ca²⁺ release media containing or lacking adenine nucleotide. The last row of table 2 shows that calmodulin reduced the Ca²⁺-induced Ca²⁺ release rate by a factor of about two. In addition, a number of drugs have been found to interact directly with the channel (table 1) [13].

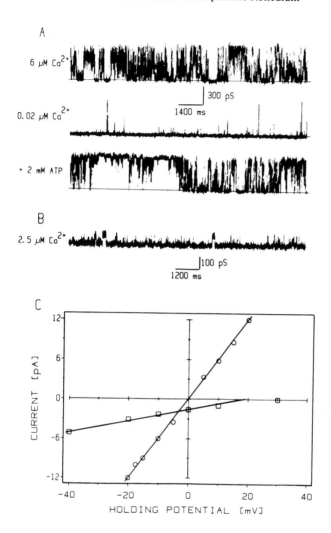

Fig. 2. Reconstitution of 30S Ca²⁺ release channel complex into planar lipid bilayers. *A* Single channel Na⁺ currents, shown as upward deflections, were recorded in symmetrical 0.5 *M* NaCl with 6 μ*M* free Ca²⁺ cis (upper trace), 0.02 μ*M* free Ca²⁺ cis (second trace), and 0.02 μ*M* free Ca²⁺ plus 2 m*M* ATP cis (third trace). *B* Single channel currents recorded after perfusion with 50 m*M* Ca Hepes trans and 125 m*M* Tris/Hepes, 2.5 μ*M* free Ca²⁺ cis. *C* Current voltage relationships for the Na⁺ (○) and Ca²⁺ (□) conducting channel. v_{Na} = 595 pS, v_{Ca} = 91 pS. Taken with permission from Lai et al. [9].

Table 2. Ca^{2+} release properties of skeletal Ca^{2+} release vesicles

Additions to release medium	$^{45}Ca^{2+}$ efflux
	$k_1 (s^{-1})$
$10^{-8} M Ca^{2+}$	0.1
$10^{-8} M Ca^{2+}, 5 \times 10^{-3} M Mg^{2+}$	0.004
$10^{-8} M Ca^{2+}, 5 \times 10^{-3} M ATP$	24
$4 \times 10^{-6} M Ca^{2+}$	1.5
$2 \times 10^{-6} M Ca^{2+}, 5 \times 10^{-3} M AMP\text{-}PCP$	90
$4 \times 10^{-6} M Ca^{2+}, 5 \times 20^{-3} M Mg^{2+}, 5 \times 10^{-3} M AMP\text{-}PCP$	14
$4 \times 10^{-6} M Ca^{2+}, 5 \times 10^{-3} M Mg^{2+}$	0.004
$4 \times 10^{-6} M Ca^{2+}, 2 \times 10^{-6} M calmodulin$	0.6

First-order rate constants of Ca^{2+} release were determined by passive loading of heavy SR vesicles with 5 mM $^{45}Ca^{2+}$ and measurements of $^{45}Ca^{2+}$ efflux rates with the use of a rapid-mixing quench-filtration protocol. Taken with permission from Meissner et al. [12].

These include caffeine which activates the purified channel in the millimolar concentration range [15]. Another interesting compound is ryanodine, a plant alkaloid. The drug binds with high and low affinity to the SR Ca^{2+} release channel, and in single channel recordings, low concentrations of ryanodine induce the formation of a long-lived open channel subconductance state whereas higher concentrations completely close the channel [9].

We have attempted to mimic the ionic conditions in relaxed and contracted muscle by placing vesicles into media containing varying concentrations of free Ca^{2+} in the presence or absence of 5 mM MgAMP-PCP (fig. 3). In these studies, we used the nonhydrolyzable ATP analog AMP-PCP to prevent the reuptake of the released $^{45}Ca^{2+}$ by the SR Ca^{2+} pump. In the absence of Mg^{2+} and nucleotide, a biphasic Ca^{2+} activation curve was obtained with Ca^{2+} release being maximal in the micromolar Ca^{2+} concentration range. A biphasic curve like the one shown in figure 3 can be most easily explained by assuming that the skeletal channel has high-affinity, Ca^{2+}-activating and low-affinity, Ca^{2+}-inhibiting binding sites. Addition of Mg^{2+} and nucleotide affected the rate and Ca^{2+} sensitivity of Ca^{2+} release. Dependence of the release rate on Mg^{2+} and nucleotide is of interest for two reasons. First, it suggests that the two ligands play an important role in the regulation of the channel. Second, local

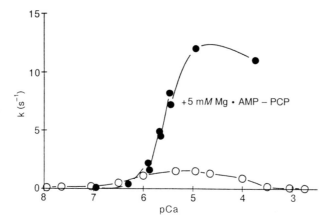

Fig. 3. Dependence of first order rate constants of $^{45}Ca^{2+}$ efflux on Ca^{2+} in the presence and absence of Mg^{2+} and AMP-PCP. SR vesicles were passively loaded with 5 mM $^{45}Ca^{2+}$ and diluted into media containing the indicated concentrations of free Ca^{2+} and zero (○) or 5 mM AMP-PCP plus 5 mM Mg^{2+} (●). In modified form from Meissner et al. [12].

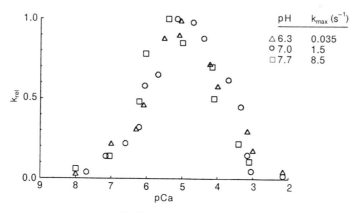

Fig. 4. Dependence of $^{45}Ca^{2+}$ efflux rate on Ca^{2+} concentration at three different pH values. SR vesicles were passively loaded with 1 mM $^{45}Ca^{2+}$ and diluted into release media containing the indicated concentrations of free Ca^{2+} at pH 6.3 (△), pH 7 (○) and pH 7.7 (□). $^{45}Ca^{2+}$ efflux rates were determined with the use of a rapid mixing-quench apparatus and by filtration [12].

changes in ATP concentration, which it is conceivable could occur during muscle exercise and fatigue, could be expected to affect the SR Ca^{2+} release process.

Another interesting finding directly related to muscle exercise and fatigue is that the Ca^{2+} release rate is strongly dependent on pH. Earlier studies have indicated that intracellular H^+ concentration affects muscle activity at different stages in the activation cycle [2, 5, 7]. In intact and skinned skeletal and cardiac muscle fiber preparations, acidosis depressed maximum tension and ATPase activity. In ion flux-vesicle and single channel measurements, SR Ca^{2+} release channel activity was greatly decreased when the pH was reduced to 6.5 and less [10, 11]. Figure 4 shows the dependence of the $^{45}Ca^{2+}$ release rate from SR vesicles on free extravesicular Ca^{2+} concentration at three different pH values [Meissner, unpubl. studies]. Ca^{2+} release rates were greatly reduced by decreasing the pH from 7.7 to 7.0 and 6.3, without an apparent change in the Ca^{2+} binding affinity of the high-affinity, Ca^{2+}-activating and low-affinity, Ca^{2+}-inhibitory binding sites of the Ca^{2+} release channel.

Conclusions

An important step towards the detailed understanding of the mechanism of excitation-contraction coupling has been the recent identification, isolation and preliminary structural and functional characterization of the high-conductance Ca^{2+} release channel in the sarcoplasmic reticulum membrane. The channel is composed of four $\sim 560,000$ polypeptides and is regulated in a complex and not yet well understood manner by several endogenous ligands including Ca^{2+}, Mg^{2+}, ATP, calmodulin, and pH. In addition, under physiological conditions, SR Ca^{2+} release activity is tightly coupled to T tubule voltage and charge movement in skeletal muscle [14]. With a few exceptions [6], such a functional interaction between T system and SR membranes remains to be demonstrated in isolated membrane fractions.

References

1   Caswell AH, Brandt NR: Does muscle activation occur by direct mechanical coupling of transverse tubules to sarcoplasmic reticulum? J Bioenerg Biomembr 1989;21:149–162.
2   Fabiato A, Fabiato F: Effects of pH on the myofilaments and the sarcoplasmic reticulum on skinned cells from cardiac and skeletal muscles. J Physiol 1978;276: 233–255.

3 Fleischer S, Inui M: Biochemistry and biophysics of excitation-contraction coupling. Ann Rev Biophys Chem 1989;18:333–364.

4 Franzini-Armstrong C: Structure of sarcoplasmic reticulum. Fed Proc 1980;39: 2403–2409.

5 Gonzales-Serratos H, Borrero LM, Franzini-Armstrong C: Changes of mitochondria and intracellular pH during muscle fatigue. Fed Proc 1974;33:1401.

6 Ikemoto N, Ronjat M, Meszaros LG: Kinetic analysis of excitation-contraction coupling. J Bioenerg Biomembr 1989;21:247–266.

7 Kentish J, Nayler WG: Effect of pH on the Ca^{2+}-dependent ATPase of rabbit cardiac and white skeletal myofibrils. J Physiol 1977;265:18p–19p.

8 Lai FA, Meissner G: The muscle ryanodine receptor and its intrinsic Ca^{2+} channel activity. J Bioenerg Biomembr 1989;21:227–245.

9 Lai FA, Misra M, Xu L, Smith HA, Meissner G: The ryanodine receptor-Ca^{2+} release channel complex of skeletal muscle sarcoplasmic reticulum. Evidence for a cooperatively coupled, negatively charged homotetramer. J Biol Chem 1989;264:16776–16785.

10 Ma J, Fill M, Knudson M, Campbell KP, Coronado R: Ryanodine receptor of skeletal muscle is a gap junction-type channel. Science 1988;242:99–101.

11 Meissner G: Adenine nucleotide stimulation of Ca^{2+}-induced Ca^{2+} release in sarcoplasmic reticulum. J Biol Chem 1984;259:2365–2374.

12 Meissner G, Darling E, Eveleth J: Kinetics of rapid Ca^{2+} release by sarcoplasmic reticulum. Effects of Ca^{2+}, Mg^{2+}, and adenine nucleotides. Biochem 1986;25:236–244.

13 Palade P, Detbarn C, Brunder D, Stein P, Hals G: Pharmacology of calcium release from sarcoplasmic reticulum. J Bioenerg Biomembr 1989;21:295–320.

14 Rios E, Pizarro G: Voltage sensors and calcium channels of excitation-contraction coupling. NIPS 1988;3:223–227.

15 Rousseau E, LaDine J, Liu QY, Meissner G: Activation of the Ca^{2+} release channel of skeletal muscle sarcoplasmic reticulum by caffeine and related compounds. Arch Biochem Biophys 1988;267:75–86.

16 Schneider MF, Chandler WK: Voltage dependent charge movement in skeletal muscle: A possible step in excitation-contraction coupling. Nature 1973;242:244–246.

17 Smith JS, Coronado R, Meissner G: Single channel measurements of the calcium release channel from skeletal muscle sarcoplasmic reticulum. J Gen Physiol 1986; 88;573–588.

18 Somlyo AV: Bridging structures spanning the junctional triad of skeletal muscle. J Cell Biol 1979;80:743–750.

19 Takeshima H, Nishimura S, Matsumoto T, Ishida H, Kangawa K, Minamino N, Matsuo H, Ueda M, Hanaoka M, Hirose T, Numa S: Primary structure and expression from complementary DNA of skeletal muscle ryanodine receptor. Nature 1989;339:439–445.

20 Zorzato F, Fujii J, Otsu K, Phillips M, Green NM, Lai FA, Meissner G, MacLennan DH: Molecular cloning of cDNA encoding human and rabbit forms of the Ca^{2+} release channel (ryanodine receptor) of skeletal muscle sarcoplasmic reticulum. J Biol Chem 1990;265:2244–2256.

Dr. Annegret Herrmann-Frank, Department of Cell Physiology, Ruhr-Universität Bochum, D-W–4630 Bochum 1 (FRG)

Marconnet P, Komi PV, Saltin B, Sejersted OM (eds): Muscle Fatigue Mechanisms in Exercise and Training. Med Sport Sci. Basel, Karger, 1992, vol 34, pp 20–42

The Contractile Performance of Normal and Fatigued Skeletal Muscle

K.A.P. Edman

Department of Pharmacology, University of Lund, Sweden

Skeletal muscle is a quite heterogeneous organ being composed of numerous separate entities, or body muscles, which differ a great deal from one another with respect to their intrinsic mechanical properties. These differences in contractile performance between muscles will not be considered at this point. Instead I will focus attention on some basic features of the contractile process that apply to skeletal muscle in general. It is now well accepted that muscle contraction is based on interaction between the two sets of interdigitating filaments, the actin and myosin filaments, which make up the sarcomere units along the fiber [Huxley, 1953; Huxley and Niedergerke, 1954; Huxley and Hanson, 1954]. For a study of the basic mechanisms of muscle contraction, it is therefore essential to turn to the single muscle fiber preparation as this enables one to control sarcomere length and, therefore, the degree of overlap between the two sets of filaments. With the techniques now available it has indeed become possible to study both force production and movement in great detail in very short segments of a single isolated muscle fiber.

A Method of Measuring Contractile Performance in Discrete Segments of Intact Single Muscle Fiber

An experimental approach for studying the contractile behavior in discrete segments of intact muscle fibers is illustrated in figure 1. For a detailed description of the technique, see Edman and Reggiani [1984]. A single fiber from the anterior tibialis muscle of the frog is mounted horizontally in a temperature-controlled bath (1–3 °C) between a force transducer

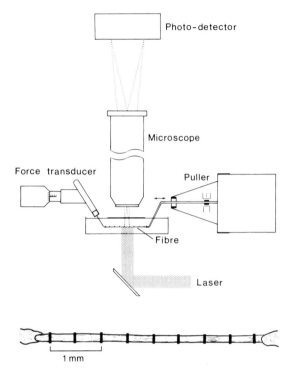

Fig. 1. Schematic illustration of apparatus used for studying the mechanical performance of marked segments along a single muscle fiber. The enlarged drawing of the fiber (bottom) illustrates the placement of the markers. The apparatus is not drawn to scale.

and an arm extending from an electromagnetic puller. The fiber is stimulated supramaximally by two platinum plate electrodes placed on either side of the preparation (not illustrated in the figure). Opaque markers are placed at 0.5-mm intervals on the upper surface of the fiber. The preparation is illuminated from below by a He-Ne laser, and an enlarged image of the fiber is projected through a microscope onto a photodiode array. The position of two adjacent markers can be sensed in this way, and the distance between the two markers, i.e. the length of one segment, can be determined. The accuracy of this measurement is better than 0.1% of the segment's length (time resolution 40 μs). The signal from a given segment can be used for feedback control of the electromagnetic puller. In this way it is possible to adjust the overall length of the fiber in such a manner that the length of that particular segment is held precisely constant throughout a contraction period.

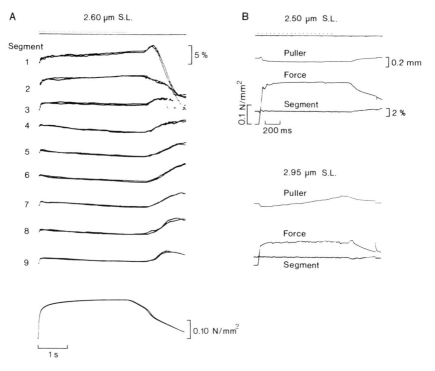

Fig. 2. Oscilloscope records illustrating standard 'isometric' recording of fiber as a whole (fixed fiber ends) *(A)* and length-clamp recording of individual fiber segment *(B).* Upper row of traces in *A* illustrate length changes of individual segments along the fiber during fixed-end tetanus. Upward deflexion of puller signal (overall fiber length) and of segment length recording indicates shortening. Note different behavior of the individual segments, and tension creep, during standard recording *(A).* By contrast, tension creep is absent when a segment is held at constant length during contraction *(B).* From Edman and Reggiani [1984].

The need for controlling sarcomere length during force recording is illustrated in figure 2. The left panel shows a standard recording from the fiber as a whole, i.e. under conditions when the overall length of the fiber is held constant (fixed fiber ends). It is evident that the various segments along the fiber do not stay constant in length during the tetanus; some segments can be seen to shorten at the expense of others, which are being stretched. Furthermore, tension does not form a distinct plateau under these conditions. Thus, after an initial rapid phase there is a slow rise of tension during the tetanus. This slow climb of force (tension creep) occurs above slack length because the stronger segments, which shorten, will steadily improve their strength as they approach the maximum of the length-tension relation. These segments will cause a corresponding force enhancement of the weaker segments which are elongated. As a result, tension will continue to rise slowly [see further Edman and Reggiani, 1984].

The situation is quite different when force is recorded while the length of a short segment is held constant during contraction (right-hand panel, fig. 2). The tension recorded under these conditions, i.e. the isometric force produced by the clamped segment, is quite constant with no creep. Even at very long sarcomere lengths no creep is seen. The tension derived by this measurement is likely to be very close to the true sarcomere isometric force in the living intact muscle fiber.

The Relations between Sarcomere Length, Force and
Velocity of Shortening

It is well established for both cardiac and skeletal muscle that the sarcomere length is an important determinant of muscle activity. However, the precise shape of the length-tension relation has remained a matter of controversy over the years due to the technical difficulties involved in assessing this relationship [Pollack and Sugi, 1984]. The main obstacle to overcome in studying this problem is the occurrence of tension creep resulting from nonuniform sarcomere behavior along the muscle fiber during contraction (see above). Figure 3 shows the well-known length-tension curve derived by Gordon, Huxley and Julian in 1966. Their curve was based on measurements that contained a substantial amount of tension creep. However, the contribution from tension creep was accounted for in the final analysis of the data. The length-tension curve derived by Gordon et al. [1966] has a polygonal shape that seemed to fit exceedingly well with the sliding-filament hypothesis. As illustrated in figure 3 only a small tension is produced at 3.65 μm sarcomere length where the thick and thin filaments were thought to be in the end-to-end position. Isometric force rises linearly as the sarcomere length is reduced to 2.2 μm at which point a maximum number of myosin cross-bridges were expected to interact with the thin filaments. Tension remains constant between 2.2 and 2.0 μm sarcomere length which seemed to be explained on the assumption that the thin filaments here move across the central zone of the thick filaments where no further cross-bridges can be formed.

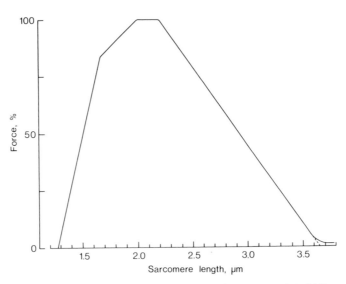

Fig. 3. The length-tension curve described by Gordon et al. [1966].

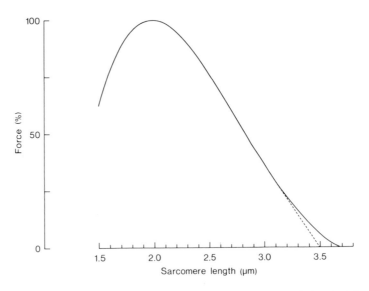

Fig. 4. The length-tension curve described by Edman and Reggiani [1987]. The relationship is based on length-clamp recordings from short (about 0.5 mm) marked segments of intact muscle fibers as described in the text.

Recent results derived from measurements on short, length-clamped segments of intact muscle fibers (see above) add further details to the length-tension relation. The length-tension curve shown in figure 4 is based on 398 measurements from 60 length-clamped segments of 20 fibers. It is clear that the sarcomere length-tension relation does not exhibit a distinct plateau. Maximum force is attained at a sarcomere length close to 2.0 μm. Above this length there is a gradual decrease in force. Note that the descending limb is quite symmetrical with a relatively straight portion in the middle and two rather similar bends at each end. The middle, straight part of the descending limb extrapolates to zero tension at 3.49 μm sarcomere length.

The length-tension curve shown in figure 4 provides some relevant information about the myofilament system. If it is assumed that force is proportional to the degree of overlap between the thick and thin filaments, it is possible to calculate the functional lengths of the two filaments from the descending limb of the curve and also the *variation* in overlap between the two filaments [for further information, see Edman and Reggiani, 1987]. The results of such an analysis are shown in figure 5. Here the solid line is the experimental curve, the same as shown in figure 4. The dots show the expected length-tension relation if the thick and thin filaments have an average functional length of 1.55 and 1.94 μm, respectively, and the extent of overlap between the two filaments is ±0.21 μm. As can be seen, these values provide a very good fit to the experimental curve.

The filament lengths arrived at in this analysis are in very good agreement with the more recent electron-microscopic measurements reported by Page [1968] and Huxley [1973]. These values of the filament lengths accord well with the finding that the middle (straight) portion of the descending limb extrapolates to the abscissa at 3.49 μm sarcomere length (fig. 4) for this is the point where the majority of the thick and thin filaments may be presumed to be in end-to-end position. The variation of filament overlap accounts for the symmetrical splays at the upper and lower ends of the descending limb. The variability in filament overlap may involve variation in length of the thin filaments or a longitudinal misalignment of the thick and thin filaments, or, a combination of these two factors [Edman and Reggiani, 1987]. Variation in thick-filament length, on the other hand, is not by itself sufficient to explain the sigmoid shape of the descending limb; this source of variability in filament overlap will only affect the lower portion of the descending limb.

Another important aspect of the contactile process is the muscle's capacity to shorten actively, for this measurement contains information

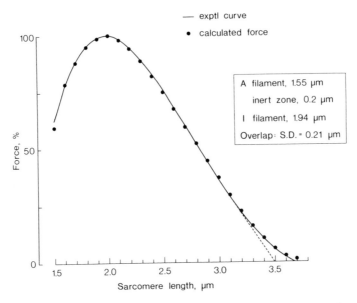

Fig. 5. Simulation of length-tension relation in terms of the sliding-filament theory using computer model described by Edman and Reggiani [1987]. Continuous line = experimental curve (the same as shown in fig. 4); filled circles = calculated values based on the following assumptions: A-filament length, 1.55 μm; central bare zone of A-filament, 0.15 μm; I-filament length, 1.94 μm; variability of overlap between A- and I-filaments, ±0.21 μm (SD). For further assumptions concerning ascending limb, see Edman and Reggiani [1987].

about the maximum turnover rate of the myosin cross-bridges. According to the sliding-filament theory, maximum velocity of shortening is determined by the rate at which the cross-bridges dissociate from the thin filament [Huxley, 1957]. Maximum velocity is reached when there is an equal number of bridges in pulling and braking positions. The maximum speed of shortening should therefore be independent of the actual number of cross-bridges that are available for interaction with the thin filament. This prediction has been tested in experiments on single muscle fibers using a method [Edman, 1979] that enables recording of the speed of shortening at zero external load. This velocity is referred to as V_0.

Figure 6 illustrates V_0 measurements at various sarcomere lengths in four different fibers. It can be seen that V_0 is quite constant between 1.7 and about 2.7 μm sarcomere length. Below 1.65 μm there is a sharp drop of V_0.

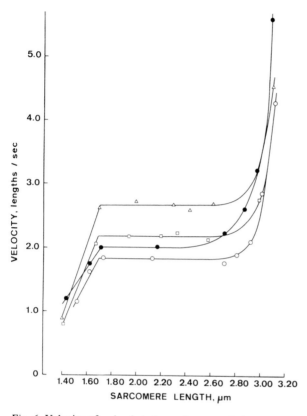

Fig. 6. Velocity of unloaded shortening recorded at different sarcomere lengths in four single muscle fibers. From Edman [1979].

Above 2.7 μm sarcomere length, on the other hand, the velocity rises steeply. V_0 is thus found to be constant over a range where the active force varies considerably (cf. fig. 4). The decrease in velocity below 1.70 μm sarcomere length can be attributed to the fact that the tips of the thick filaments here come up against the Z disk. This will create a braking force directed against the sliding movement. The rise in velocity at long lengths can be related to the passive tension that exists at these lengths. The passive tension which is built up when the sarcomeres are prestretched to these lengths will act in parallel with the contractile unit. It will tend to shorten the sarcomeres and therefore cause the filaments to slide at a higher speed than they are able to achieve on their own. The results presented in figure 6 clearly show, however, that under conditions when there are no passive forces interfering with the sliding movement, i.e. between 1.7 and about 2.7 µm sarcomere length, *the velocity of shortening is independent of the extent of filament overlap.* V_0 is therefore independent of the number of myosin bridges that are able to interact with the thin filament, a finding that fully supports the sliding filament model and the idea that myosin cross-bridges act as independent force generators [Huxley, 1957].

Deactivation of Contractile System by Active Shortening

Striated muscle that is allowed to shorten during activity temporarily loses some of its capacity to produce force. This depressant effect of shortening has been demonstrated in both isolated muscle preparations [Jewell and Wilkie, 1960; Edman and Kiessling, 1971; Briden and Alpert, 1972] and in muscles in situ in the body [Joyce et al., 1969]. A detailed study of this phenomenon has previously been performed on isolated muscle fibers [Edman, 1975, 1980, 1981; Ekelund and Edman, 1982]. Some characteristic features of the 'movement effect' are considered here.

Figure 7 demonstrates the depressant effect of active shortening in a frog single muscle fiber. The upper record (A) shows a partially fused tetanus performed at 2.05 µm sarcomere length. At the stimulus frequency used, the individual twitch periods show up as distinct humps. In the lower record (B) the same stimulation is repeated, but in this case the contraction is started at a longer sarcomere length, 2.55 µm, and the fiber is allowed to shorten to 2.05 µm during the first twitch period. The fiber is arrested at this length to develop isometric force as in the first myogram. It can be clearly seen, by comparing the two myograms, that force development is markedly depressed after the shortening phase. Twitch No. 2 in myogram B is thus considerably lower than the first twitch in the control run. This is of relevance since tension starts from the zero level and occurs at the same sarcomere length in both cases. The reduced amplitude of twitch period No. 2 in the lower myogram thus represents a true depression of the fiber's capacity to produce force, due to the preceding shortening. It is evident, however, that the depressant effect of shortening disappears with time during continued stimulation. The fiber has completely regained its strength 1–2 s after the end of the shortening phase.

The evidence suggests that shortening reduces the degree of activation of the contractile system. This is indicated by the fact (fig. 8) that the depressant

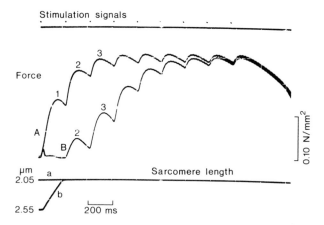

Fig. 7. Depressant effect of shortening during a partially fused tetanus of single muscle fiber. In both records force is developed at 2.05 μm sarcomere length. In myogram B contraction is initiated at 2.55 μm sarcomere length and the fiber is allowed to shorten to 2.05 μm during the first twitch period. The first few twitch periods are numbered for identification in the text. Note marked depression of the fiber's ability to produce force after shortening in myogram B.

effect of shortening is diminished if the state of activity in the fiber is increased. The top left panel *(A)* of figure 8 illustrates the depressant effect during a partially fused tetanus. Myogram *a* (control) shows the redevelopment of force after a small shortening which is just sufficient to produce a drop in tension to zero and an immediate redevelopment of force. In myogram *b* the amplitude of shortening is larger, and the release is timed appropriately so as to make the redevelopment of force start at the same time as in the control. The extra amount of shortening produced in myogram *b* can be seen to depress the amplitude of the redeveloped force by approximately 25%. Panel *D* shows, for comparison, the corresponding effects of shortening during a *fused* tetanus in the same fiber, i.e. under conditions when the contractile system can be presumed to be fully activated. In this case the shortening is produced during the plateau of the tetanus, before the last stimulus. The depressant effect (approximately 7%) can now be seen to be considerably smaller than during the partially fused tetanus.

In the middle panels *(B* and *E)* of figure 8 the same movements are performed after addition of caffeine in a concentration that is known to increase the release of calcium from the sarcoplasmic reticulum [Blinks et al., 1978]. The presence of caffeine greatly reduces the depressant effect of

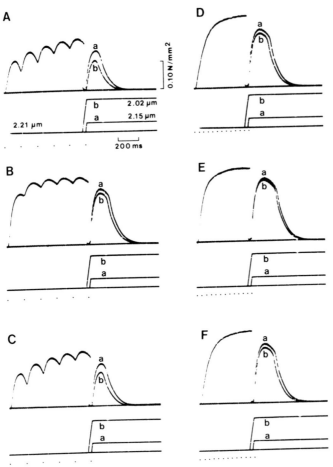

Fig. 8. Influence of caffeine on the depressant effect of active shortening. Control (a) and test (b) movements performed during the last cycle of partially fused tetanus *(A–C)* and prior to the last stimulus of fused tetanus *(D–F)*. *A, D* Fiber immersed in ordinary Ringer's solution. *B,E* 45–65 min after addition of 0.5 mM caffeine to bathing fluid. *C, F* 50–65 min after change to ordinary Ringer's solution. Lower traces indicate sarcomere length before and after the shortening. From Edman [1980].

shortening, both in absolute and relative terms, during the partially fused tetanus, and the movement effect is nearly abolished by caffeine during the fused tetanus. Removal of caffeine (panels *C* and *F*) restores the depressant effect completely. The results support the view that active shortening temporarily reduces the state of activation of the contractile system, most likely

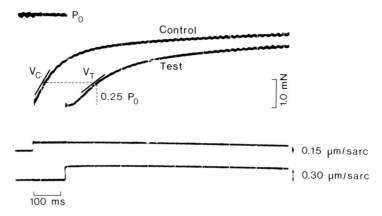

Fig. 9. Depressant effect of active shortening in frog skinned muscle fiber. Two superimposed records illustrating redevelopment of force after small (control) and large (test) shortening performed during calcium-induced contracture of skinned fiber. Lower traces show shortening steps calibrated in μm/sarcomere. V_C and V_T denote rate of force development (here used as an index of the contractile strength) at 25% of maximum isometric force. Redevelopment of force recorded between 2.30 and 2.00 μm sarcomere length, i.e. close to maximum of length-tension relation. From Ekelund and Edman [1982].

by reducing the affinity for calcium at the troponin binding-sites. This is in line with the observation that the movement effect is large when the contractile system is submaximally activated as during a twitch or a partially fused tetanus. This interpretation of the movement effect is also strongly supported by the finding that there is virtually no depression by shortening when the myofilament system is maximally, or supramaximally, activated as is the case during a fused tetanus in the presence of caffeine. Under these conditions a decrease in calcium affinity at the binding sites would be compensated by a higher free calcium concentration in the myofibrillar space.

Further evidence supporting the view that the movement effect is based on a change in the myofilament system is provided by the finding that the depressant effect of shortening also appears in *skinned* muscle fibers. These are preparations in which the membrane function has been abolished either by removing the membrane mechanically or by making the membrane leaky by treatment with glycerol and detergent. Such fibers can be activated to different degrees by varying the free calcium concentration in the surrounding medium. Figure 9 shows results from such a preparation. Illustrated are

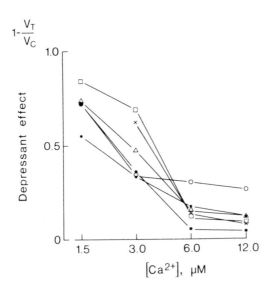

Fig. 10. Depressant effect of active shortening of skinned muscle fibers from the mouse related to the free calcium concentration. Measurement of depressant effect performed as in figure 9. Each set of data denoted by the same symbol is from a single preparation. From Ekelund and Edman [1982].

two superimposed records on a relatively fast time base. In both records the fiber has first developed maximum tension (P_0), and the fiber is thereafter released to shorten and redevelop force. In the control myogram the shortening is small (0.15 μm/sarcomere) whereas in the test myogram the fiber is allowed to shorten over a longer distance (0.30 μm/sarcomere). The release steps produced by the puller (lower traces) are fast enough to slacken the fiber which means that the fiber shortens at maximum speed. It can be seen that the *rate* of force development (which is used here as an index of the contractile strength) is considerably lower after the larger movement indicating that active shortening also depresses force production in the skinned fiber.

Figure 10 shows that the depressant effect of shortening is critically dependent on the free calcium concentration around the myofibrils. These are results from skinned *mammalian* (mouse) muscle fibers. As can be seen, the depressant effect is inversely related to the free calcium concentration. Thus, the movement effect is large when the calcium concentration is low. Conversely, the depressant effect of shortening almost disappears when the calcium concentration is raised to levels which are maximal, or supramaximal, for activation of the contractile system.

It is reasonable to conclude from the above findings that active sliding of the filaments leads to a true deactivation of the contractile system. Activity is needed for the depressant effect to appear, i.e. the two filaments have to interact during sliding for the movement effect to occur [Edman, 1980]. This supports the view [Edman, 1975, 1980; Ekelund and Edman, 1982] that the myosin bridges somehow interfere with the regulatory proteins as the bridges go through activity cycles along the thin filament. The movement effect may in fact be a manifestation of a cooperative activity between the proteins of the calcium-regulated thin filament leading to a transitory decrease in calcium affinity and, therefore, to a transient deactivation of the contractile machinery. All the evidence suggests that the movement effect is an integral part of the sliding-filament process, and it is reasonable to suppose that it does play a part in normal life. Mammalian motor units are activated to produce partially fused tetani, and under such conditions the movement effect would be pronounced (cf. fig. 7). The effect is short-lasting, however, and the fibers regain their full contractile strength within 1–2 s during continued stimulation [see further Edman, 1975; Ekelund and Edman, 1982]. The rationale of having this effect operating during muscle activity is unclear at the present time. It may serve as a safety mechanism to prevent overuse of the muscles.

Cellular Mechanisms of Muscle Fatigue

There is an obvious difficulty in reproducing human or animal fatigue in an isolated muscle preparation as there is no clear information on how extensively a muscle can be stimulated by its nerve in situ in the body [Edwards, 1981]. It is easy enough, of course, to stimulate an isolated preparation until some aspect of the excitation mechanism fails, but it remains uncertain if this stage is ever reached in vivo.

In the present study of fatigue, performed on frog isolated muscle fibers, a stimulation protocol has been used that leads to 25–30% depression of tetanic force without causing any obvious failure of the excitation-contraction coupling. In the control series the fiber is stimulated to produce a 1-second isometric tetanus at regular 5- or 15-min intervals. Fatigue is produced by reducing the resting intervals between tetani to 15 s. As illustrated in figure 11, the peak isometric force is reduced to about 70–75% of the

Tetanic force

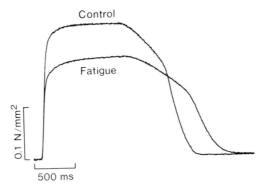

Fig. 11. Superimposed isometric tetani of a single muscle fiber recorded after attainment of steady state at two contraction frequencies. Interval between contractions: 5 min (control) and 15 s (fatigue).

control value by this procedure. The records in figure 11 also illustrate that fatiguing stimulation leads to reduced rate of rise of force and a marked slowing of the relaxation phase.

The decrease in force during fatigue could have two causes. There may be a decrease in the number of cross-bridges formed or the force produced by each individual bridge may be reduced. In an attempt to elucidate this point we have studied the stiffness of the fiber as fatigue develops. The instantaneous stiffness of the contractile system (measured by imposing very fast, small length changes) is generally considered to be a measure of the *number* of cross-bridges that are attached to the thin filaments [Ford et al., 1977]. A simultaneous measurement of active force and stiffness during development of fatigue therefore offers a possibility to evaluate whether or not a change in the number of attached cross-bridges is alone responsible for the decrease in force.

Figure 12 illustrates the approach used for measuring stiffness [see further, Edman and Lou, 1990]. A 4-kHz length oscillation of constant amplitude is applied to one end of the fiber throughout the tetanus period. For the purpose of illustration the oscillation shown in figure 12 is cut short during the plateau phase. The amplitude of the length oscillation corresponds to approximately 1.7 nm/half sarcomere, which is only a fraction of the estimated compliance of a cross-bridge during a working cycle [Ford et al.,

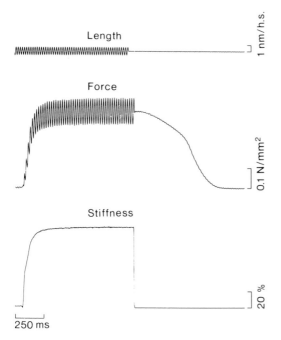

Fig. 12. Oscilloscope records illustrating the approach used for measuring fiber stiffness by means of sinusoidal length oscillation. Upper trace = 4 kHz length oscillation of fiber (puller movement) calibrated in nm/half sarcomere. For the purpose of demonstration the length oscillation is stopped on the tetanus plateau. Middle trace = Tetanic force with superimposed response to length perturbation. Lower trace = Amplitude of oscillatory force response, 'stiffness signal' (obtained by passing force signal through a narrow band-pass filter (4 kHz) followed by rectification).

1977]. The imposed length oscillation gives rise to corresponding changes in fiber tension, the amplitude of this tension oscillation being proportional to the actual stiffness of the fiber. The amplitude of force oscillation is determined on-line by filtration and rectification of the force signal [see further, Edman and Lou, 1990]. A moment-to-moment read-out of the stiffness is obtained in this way (bottom record in figure 12).

Figure 13 shows how stiffness is affected, relative to force, during development of fatigue. The left-hand records are force myograms, the same as shown in figure 11. The right-hand traces illustrate the corresponding changes in stiffness. Traces 1 are control records and traces 2 recordings after development of fatigue. It can be seen that stiffness undergoes a smaller change during fatigue than tension does. The diagram shows the relative

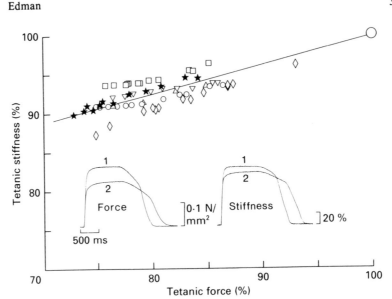

Fig. 13. Relation between maximum tetanic force and maximum tetanic stiffness during development of fatigue in six single muscle fibers. Data expressed as percentage of maximum force and maximum stiffness recorded under control conditions in respective fiber. Values from a given fiber are denoted by the same symbol. Least-squares regression of stiffness (S_{tet}) upon force (F_{tet}) provides the following relation (line): $S_{tet} = 0.369 \, F_{tet} + 62.91$. Insets: Superimposed records of tetanic force (left) and tetanic stiffness (right) under control conditions (trace 1) and after fatiguing stimulation (trace 2). From Edman and Lou [1990].

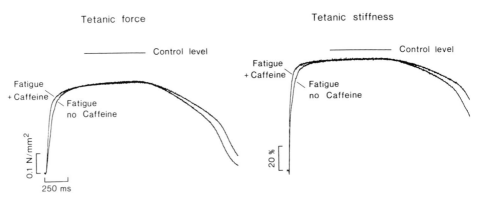

Fig. 14. Superimposed records of tetanic force (left) and tetanic stiffness (right) in the presence and absence of caffeine (0.5 mM) after fatigue had developed in a single muscle fiber. The control ('rested state') levels of force and stiffness indicated by horizontal line above the records. Note that caffeine increases the rate of rise of both force and stiffness but does not affect the total amplitude of either measurement.

changes in maximum tetanic force and maximum tetanic stiffness during development of fatigue in six fibers. The rested-state values of force and stiffness are here expressed as 100% in each case. The abscissa shows the decrease in force as the fibers go from rest to fatigue, whereas the ordinate indicates the corresponding change in stiffness. A regression analysis based on these data shows that a 25% depression of tetanic force is associated with merely 9% reduction in fiber stiffness. It can be concluded from these findings that there is a slight decrease in the number of active cross-bridges during fatigue and this accounts for part of the force reduction. However, the most important cause of the tension decline during fatigue is reduced ability of the individual cross-bridge to produce active force.

The slight decrease in number of active cross-bridges could mean that the fiber becomes submaximally activated in the fatigued state. Figure 14 presents evidence against this view. In this experiment caffeine has been added to the fiber in a twitch potentiating concentration *after* fatigue has developed. The superimposed records show force and stiffness in the presence and absence of caffeine after the fiber has been fatigued. Caffeine can be seen to increase the initial rate of rise of both force and stiffness suggesting that caffeine did enhance the release of activator calcium. It is clear, however, that caffeine does not affect maximum tetanic force, nor maximum tetanic stiffness, in the fatigued state. This means that the contractile system was maximally activated even after the fiber had been fatigued. Taken together these results suggest strongly that fatigue, to the degree studied here, is based entirely on altered kinetics of cross-bridge function with no significant change in the state of activation of the contractile system.

The altered cross-bridge function during fatigue does not merely affect the fiber's ability to produce force. It also leads to a decrease in the maximum speed of shortening of the fiber as is demonstrated in figure 15. The results show how maximum speed of shortening is affected, relative to force, as individual muscle fibers go into fatigue. It can be seen that the velocity is unaffected as long as the tetanic force is reduced by less than approximately 10%. However, with increasing fatigue and further depression of the isometric force the maximum speed of shortening is also progressively reduced. This means that not only are the bridges less capable of producing force in the fatigued state (see above), the speed at which the bridges are able to go through a cycle of activity is also diminished.

Fatiguing stimulation apparently leads to some change of the intracellular milieu that affects cross-bridge function. Accumulation of breakdown products of ATP such as ADP, inorganic phosphate and H^+ is known to occur

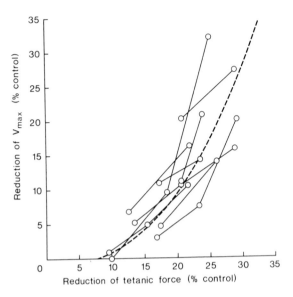

Fig. 15. Decrease in maximum velocity of shortening (ordinate) related to depression of maximum tetanic force (abscissa) during development of fatigue in single muscle fibers. Each set of data points connected by a solid line is from a single fiber. Dashed line derived by least-squares regression analysis of velocity upon force. Co-ordinates expressed as percent of V_{max} and tetanic force recorded under control (nonfatigued) conditions.

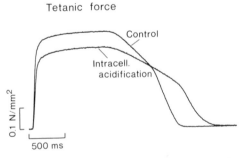

Fig. 16. Superimposed oscilloscope records of tetanic force recorded in ordinary Ringer's solution (control) and in CO_2-containing solution causing intracellular acidification. Single muscle fiber. For further information, see text.

during sustained muscle activity [Edwards et al., 1975; Dawson et al., 1978, 1980], and these products all affect the kinetic properties of the contractile system [see Edman and Lou, 1990]. The effects of lowered intracellular pH can be tested in the intact muscle fiber by increasing the CO_2 concentration of the extracellular medium [Edman and Mattiazzi, 1981; Curtin and Edman, 1989; Edman and Lou, 1990]. Such experiments show that the contractile changes during fatigue can be simulated well at single fiber level by lowering the intracellular pH (pH_i). This is exemplified in figure 16 which shows that lowered pH_i, similar to fatigue, reduces both the rate of rise of force and the total amplitude of the tetanus and causes a marked slowing of the relaxation phase. A decrease of the intracellular pH also reduces V_0, the maximum speed of shortening, and this change bears approximately the same relation to the decrease in tetanic force as does the V_0 change observed in fatigue [Edman and Mattiazzi, 1981]. It is unlikely, however, that intracellular acidification is the only cause of muscle fatigue for some of the pH effects differ quantitatively from the fatigue effects. Examples of such differences between the two interventions are given by Edman and Lou [1990]. Some other metabolic factor, or factors, are therefore likely to play a role, in addition to lowered pH_i, during muscle fatigue. Both ADP and inorganic phosphate are known to affect cross-bridge function [Chase and Kushmerick, 1988; Cooke et al., 1988; Godt and Nosek, 1989], and accumulation of these two products during fatigue may be supposed to modulate the effects produced by the intracellular acidification.

Another possible cause of muscle fatigue, which may play a part under more extreme stimulation conditions than those used here, is failure of the inward spread of activation into the muscle fiber. One way of assessing this phenomenon is to shorten the fiber below slack length to find out if the myofibrils in the center of the fiber have a straight appearance or not [Gonzalez-Serratos et al., 1981]. A muscle fiber cannot shorten below its slack length unless it is activated. Thus, if the central myofibrils are inactive, they will have a wavy appearance while the peripheral layers of the fiber, which have shortened actively, are straight.

Such experiments have recently been performed at this laboratory using two different fatiguing protocols: (1) Our standard protocol (used in the experiments reported above): 1-second tetanus every 15 s. (2) A more intense stimulation protocol: a single stimulus applied at 2- to 3-second intervals until tetanic force is reduced by approximately 50% of its control value. The latter stimulation protocol has been utilized in several previous studies [Eberstein and Sandow, 1963; Gonzalez-Serratos et al., 1978, 1981; Nassar-

Gentina et al., 1978; Lännergren and Westerblad, 1986, 1989; Allen et al., 1989]. Fibers fatigued by either of the two protocols were allowed to shorten during tetanus below slack length, to approximately 1.7 µm sarcomere length. Here the fibers were quickly frozen during activity and thereafter processed for electron microscopy.

The results of the quick-freeze experiments show [Edman and Lou, unpubl. data] that fatiguing protocol 1 does not lead to any detectable inactivation of myofibrils in any part of the fiber. The situation is quite different, however, when fatigue is produced according to protocol 2. In this case the core of the fiber exhibits a wavy appearance after the fiber has shortened below slack length indicating that the central myofibrils are not properly activated. This failure of activation may arise because potassium accumulates in the transverse tubules when the stimulation frequency is high. An increased concentration of potassium in the T tubules, possibly in combination with osmotic changes in the T tubular membrane [Gonzalez-Serratos et al., 1978], will impair the inward spread of the action potential and so make the inner parts of the fiber inaccessible for the stimulus. Whether such an extreme situation, with failure of the inward spread of activation, will ever occur in vivo is still uncertain. It is clear, however, that before inactivation starts to develop during frequent stimulation, the myofilament system has already lost a great deal of its ability to produce force and to shorten actively.

References

Allen DG, Lee JA, Westerblad H: Intracellular calcium and tension during fatigue in isolated single muscle fibres from *Xenopus laevis.* J Physiol 1989;415:433–458.

Blinks JR, Rüdel R, Taylor SR: Calcium transients in isolated amphibian skeletal muscle fibres: Detection with aequorin. J Physiol 1978;277;291–323,

Briden KL, Alpert NR: The effect of shortening on the time-course of active state decay. J Gen Physiol 1972;60:202–220.

Chase PB, Kushmerick MJ: Effects of pH on contraction of rabbit fast and slow skeletal muscle fibers. Biophys J 1988;53:935–946.

Cooke R, Franks K, Luciani GB, Pate E: The inhibition of rabbit skeletal muscle contraction by hydrogen ions and phosphate. J Physiol 1988;395:77–97.

Curtin NA, Edman KAP: Effects of fatigue and reduced intracellular pH on segment dynamics in 'isometric' relaxation of frog muscle fibres. J Physiol 1989;413;159–174.

Dawson MJ, Gadian DG, Wilkie DR: Muscular fatigue investigated by phosphorus nuclear magnetic resonance. Nature 1978;274:861–866.

Dawson MJ, Gadian DG, Wilkie DR: Mechanical relaxation rate and metabolism studied in fatiguing muscle by phosphorus nuclear magnetic resonance. J Physiol 1980;299: 465–484.

Eberstein A, Sandow A: Fatigue mechanisms in muscle fibres; in Gutman E, Hnik P (eds): The Effect of Use and Disuse on Neuromuscular Functions. Nakladatelstvi Ceskoslovenske akademie ved Praha, Prague, 1963, pp 515–526.

Edman KAP: Mechanical deactivation induced by active shortening in isolated muscle fibres of the frog. J Physiol 1975;246:255–275.

Edman KAP: The velocity of unloaded shortening and its relation to sarcomere length and isometric force in vertebrate muscle fibres. J Physiol 1979;291:143–159.

Edman KAP: Depression of mechanical performance by active shortening during twitch and tetanus of vertebrate muscle fibres. Acta Physiol Scand 1980;109:15–26.

Edman KAP, Kiessling A: The time course of the active state in relation to sarcomere length and movement studied in single skeletal muscle fibres of the frog. Acta Physiol Scand 1971;81:182–196.

Edman KAP, Lou F: Changes in force and stiffness induced by fatigue and intracellular acidification in frog muscle fibres. J Physiol 1990;424:133–149.

Edman KAP, Mattiazzi A: Effects of fatigue and altered pH on isometric force and velocity of shortening at zero load in frog muscle fibres. J Muscle Res Cell Motil 1981;2:321–334.

Edman KAP, Reggiani C: Redistribution of sarcomere length during isometric contraction of frog muscle fibres and its relation to tension creep. J Physiol 1984;351:169–198.

Edman KAP, Reggiani C: The sarcomere length-tension relation determined in short segments of intact muscle fibres of the frog. J Physiol 1987;385:709–732.

Edwards RHT: Human muscle and fatigue; in Porter R, Whelan J (eds): Ciba Foundation Symposium 82: Human Muscle Fatigue: Physiological Mechanisms. London, Pitman Medical, 1981.

Edwards RHT, Hill DK, Jones DA: Metabolic changes associated with the slowing of relaxation in fatigued mouse muscle. J Physiol 1975;251:287–301.

Ekelund M, Edman KAP: Shortening induced deactivation of skinned fibres of frog and mouse striated muscle. Acta Physiol Scand 1982;116:189–199.

Ford LE, Huxley AF, Simmons RM: Tension responses to sudden length change in stimulated frog muscle fibres near slack length. J Physiol 1977;269:441–515.

Godt RE, Nosek TM: Changes of intracellular milieu with fatigue or hypoxia depress contraction of skinned rabbit skeletal and cardiac muscles. J Physiol 1989;412:155–180.

Gonzalez-Serratos H, Somlyo AV, McClellan G, Shuman H, Borrero LM, Somlyo AP: Comparison of vacuoles and sarcoplasmic reticulum in fatigued muscle: Electron probe analysis. Proc Natl Acad Sci USA 1978;75:1329–1333.

Gonzalez-Serratos H, Garcia M, Somlyo A, Somlyo AP, McClellan G: Differential shortening of myofibrils during development of fatigue. Biophys J 1981;33:224a.

Gordon AM, Huxley AF, Julian FJ: The variation in isometric tension with sarcomere length in vertebrate muscle fibres. J Physiol 1966;184:170–192.

Huxley AF: Muscle structure and theories of contraction. Prog Biophys Biophys Chem 1957;7:255–318.

Huxley AF, Niedergerke R: Structural changes in muscle during contraction: Interference microscopy of living muscle fibres. Nature 1954;173:971–973.

Huxley HE: Electron-microscope studies of the organization of the filaments in striated muscle. Biochim Biophys Acta 1953;12:387.

Huxley HE: Molecular basis of contraction in cross-striated muscle; in Bourne G (ed): The Structure and Function of Muscle, vol 1, ed 2. New York, Academic Press, 1973.

Huxley HE, Hanson J: Changes in the cross-striations of muscle during contraction and stretch and their structural interpretation. Nature 1954;173:973–977.

Jewell BR, Wilkie DR: The mechanical properties of relaxing muscle. J Physiol 1960;152: 30–47.

Joyce GC, Rack PMH, Westbury DR: The mechanical properties of cat soleus muscle during controlled lengthening and shortening movements. J Physiol 1969;204:461–474.

Lännergren J, Westerblad H: Force and membrane potential during and after fatiguing, continuous high-frequency stimulation of single *Xenopus* muscle fibres. Acta Physiol Scand 1986;128:359–368.

Lännergren J, Westerblad H: Maximum tension and force-velocity properties of fatigued, single *Xenopus* muscle fibres studied by caffeine and high K^+. J Physiol 1989;409: 473–490.

Nassar-Gentina V, Passonneau JV, Vergara JL, Rapoport SI: Metabolic correlates of fatigue and recovery from fatigue in single frog muscle fibers. J Gen Physiol 1978;72:593–606.

Page S: Fine structure of tortoise skeletal muscle. J Physiol 1968;197:709–715.

Pollack GH, Sugi H (eds): Contractile Mechanisms in Muscle. New York, Plenum Press, 1984.

K.A.P. Edman, MD, PhD, Department of Pharmacology, University of Lund, S-223 62 Lund (Sweden)

Marconnet P, Komi PV, Saltin B, Sejersted OM (eds): Muscle Fatigue Mechanisms in
Exercise and Training. Med Sport Sci. Basel, Karger, 1992, vol 34, pp 43–53

Fatigue Mechanisms in Isolated Intact Muscle Fibers from Frog and Mouse[1]

Jan Lännergren

Department of Physiology II, Karolinska Institutet, Stockholm, Sweden

Introduction

There are various definitions of muscle fatigue. In the present context, fatigue is defined as any decline in force output during prolonged stimulation. The following overview is based on results obtained from single fibers. Work on isolated fibers has several advantages and also some disadvantages. The major advantages are that responses are all-or-none, there is no uncertainty about possibly damaged fibers; adequate oxygenation is easy to ensure; the extracellular medium can be rigorously controlled and quickly altered if desired; fibers of different types can be studied separately. For fatigue studies there are also some potential disadvantages: the preparation is usually continuously superfused with fresh salt solution and changes in the vicinity of the fiber, which may occur during normal development of fatigue in a whole muscle, and which may contribute to force loss, are eliminated. Also, the small size of the preparations makes measurements of metabolite concentrations difficult. Still, single fiber preparations can be very useful for providing information about basic fatigue mechanisms, to which possible effects of changes in the extracellular environment might then be added.

Fiber Preparations and Stimulation Schemes

Two kinds of preparations have been used: (i) frog fibers and (ii) mouse fibers. In contrast to many other studies on frog muscle the species used here

---

[1] This overview is based on research supported by the Swedish Medical Research Council (project 3642), funds at the Karolinska Institutet and the Bergvall Foundation.

is *Xenopus (laevis)*, an African aquatic frog, rather than the more commonly used *Ranas*. One advantage of *Xenopus* is that fibers of different types can be easily recognized during the dissection. Fibers for the present experiments are dissected from the lumbrical muscles of the foot, because these fibers are short (about 1.5 mm) which is advantageous, especially for microelectrode recordings. The isolated mammalian fiber is a fairly novel preparation; we dissect them (without using enzymes) from the flexor brevis muscle of the foot [1]. The mouse fibers are quite small, typically 35 μm in diameter and 800 μm in length.

Fiber Types

Mouse Fibers. The two major fiber types in mammalian muscle are familiar: fast twitch (type II) and slow twitch (type I). The mouse foot muscle we use contains predominantly fast-twitch fibers and nearly all the results are from such fibers. Whether they are of type IIA or type IIB has not yet been determined.

Xenopus Fibers. It is probably less well known that there is also a diversity of twitch fiber types in amphibian muscle. This is particularly evident in *Xenopus*: there are three major types (1, 2 and 3) which differ in myosin composition and contractile speed [2], and most importantly here, in their resistance to fatigue. As already mentioned, the different types can be identified and selected during dissection, which is helpful because fatigue mechanisms might be different in different fiber types and these can be studied separately. However, the wide variation in fatigue resistance causes a problem when designing a protocol for fatiguing stimulation, as discussed below.

Fatigue Protocols. Two kinds of fatigue protocols have been used: (A) *Continuous*, relatively *high frequency* stimulation (70 Hz). This gives a nearly smooth tetanic contraction; this scheme has been used for *Xenopus* fibers only. (B) *Intermittent* tetanic stimulation (350 or 500 ms tetani, also at 70 Hz). This has been used for the majority of the experiments, both on *Xenopus* and on mouse fibers.

With protocol B the problem of differences in fatigue resistance occurs. With a constant tetanus interval some fibers will fatigue very quickly, others will continue to give high force for half an hour or more. To solve this we have used a scheme in which the *duty cycle* (stimulation time/tetanus interval) is increased every 2 min [for details, see ref. 3]. Experiments on frog fibers have been performed at 20–22 °C and on mouse fibers at 25 °C.

Results and Discussion

Force Decline

Fatigue Protocol A: With this stimulation scheme force goes down fairly quickly with a half-time of some 6–7 s. Intracellular recordings of membrane potential and action potentials show several changes: (1) Action potentials quickly become broadened and diminish in amplitude. The early negative after-potential becomes less distinct. (2) The baseline potential gradually becomes more positive, i.e. a depolarization to about –55 mV.

The cause of tension decline with this type of stimulation is probably not changes 1 and 2 per se since depolarization of rested fibers with 14 mM K$^+$ (to give a membrane potential of about –55 mV and a 'fatigued' action potential) gives an augmented twitch [4]. Force recovery is exceedingly rapid after high-frequency stimulation and may even occur *during* stimulation if the impulse frequency is lowered, hence metabolic changes are unlikely to be responsible. It appears more likely that a step in excitation-contraction coupling (ECC) fails, probably impulse propagation down the t tubuli. Strong evidence was provided for this view by injecting fibers with the Ca indicator fura-2 and visualizing Ca release at various depths in the fiber by digital imaging fluorescence microscopy. A spatial gradient in Ca$_i$ from the membrane towards the center of the fiber was then observed [5].

Fatigue Protocol B. With this type of stimulation large differences in fatigue resistance between *Xenopus* fibers of types 1, 2 and 3 are revealed. Endurance, here measured as stimulation time for tension to fall to 0.75 P$_o$ correlates very well (r = 0.94) with the quotient of SDH activity/myofibrillar ATPase activity as determined by quantitative histochemistry [van der Laarse, Lännergren and Diegenbach, in preparation].

What is fatigue due to with this type of stimulation?

(1) Failure of t tubule propagation as with protocol A? This appears unlikely since experiments with visualization of Ca distribution show that during intermittent tetanic stimulation there is *no gradient*, rather a homogeneous decrease in Ca$_i$ [5]. If ECC failure occurs it has to appear at a later stage.

(2) Depletion of substrate for metabolism? This also seems unlikely since a second fatigue run can be obtained after adequate recovery in a medium which does not contain glucose.

(3) Inhibitory action of metabolites on cross-bridge function? The two most commonly discussed metabolites are hydrogen ions (H$^+$) and inorganic

phosphate (P_i). Intracellular acidification may play a role, but is unlikely to be the only, or even dominating factor under the present experimental conditions. Evidence for this is: (i) pH_i has been measured with ion-selective microelectrodes in *Xenopus* fibers demonstrating that for the same level of fatigue (0.4 P_o) pH_i is lowered to 6.15–6.50 for type 1 fibers but only to 6.50–6.85 for type 2 fibers [6]. (ii) Intracellular acidosis induced by exposure to 30% CO_2 in mouse fibers (calculated pH_i = 6.5–6.6) gives only moderate force depression, about 20% [7]. There are also examples from several other investigations that fatigue can develop with only minor changes in pH_i [8, 9].

There are many studies on skinned fibers which show that inorganic phosphate P_i has a force-depressing effect [10, 11]. An increase in P_i may well contribute to force depression in the present experiments, but a potential argument against increased P_i being the sole cause is the time course of the fatigue curves, especially in the mouse fibers. P_i would be expected to build up gradually, but the major force decline starts quite suddenly and then develops fairly rapidly [7].

(4) Some other form of ECC failure than failing t tubule propagation? There are two lines of evidence that this may be of major importance:

(i) Measurements of Ca_i (Ca transients) during fatiguing stimulation. Ca transients recorded in *Xenopus* fibers with aequorin show first an increase in amplitude (in force range 0.95–0.85 P_o), then a gradual decline. The amplitude of the Ca transient is about 45% of original when force is down to 0.5 P_o [12]. Similar studies with fura-2 as a Ca indicator also show a decline in Ca transients during the final fall in tension, but they also reveal a gradual increase in resting Ca_i [13]. In this study evidence was also obtained for a decreased Ca sensitivity of the contractile elements in the fatigued state.

(ii) Response to *caffeine*. We have used caffeine extensively both for *Xenopus* and mouse fibers as an analytic tool. Caffeine has at least three actions: (1) It facilitates release of Ca from the sarcoplasmic reticulum (SR) (increases open probability of Ca channels [14]); high doses can directly induce release. (2) It inhibits Ca reuptake into the SR. Actions 1 and 2 mean that the normal EC coupling can be improved or bypassed. (3) It increases the Ca sensitivity of the thin filaments [15]. For other references, see Fryer and Neering [16].

Caffeine Experiments on Xenopus *Fibers.* A relatively high concentration was used (8–15 mM), which causes a large contracture in rested fibers

(84% of P_o). When the same concentration was applied to fibers in the fatigued state (about 0.4 P_o) tension quickly rose to a similar value (mean = 81% of P_o, n = 20 [17]). This means that fatigued fibers *can* produce high force if the activation is artificially made more effective. Precisely how effective the tension restoration is in these fibers cannot be ascertained since caffeine contractures in the rested state were lower than maximum tetanic tension.

Caffeine Experiments on Mouse Fibers. Corresponding, and more extensive, experiments have been performed on mouse fibers, also using 15 mM caffeine. However, in mouse muscle, at close to room temperature, 15 mM is subthreshold, that is, it does not induce a contracture in the rested state. A slight potentiation of tetanic (70 Hz) tension is, however, observed, amounting to about 5%. In the fatigued state (0.3 P_o) application of 15 mM caffeine causes a prompt and nearly threefold increase in force output [7]. The average force level of fatigued fibers in the presence of caffeine was 82.5% of P_o (n = 10). If a comparison is made with the rested 70 Hz + caffeine tension (1.05 P_o) there is thus a component of tension loss which cannot be reversed by caffeine, amounting to about 25% of P_o, and a larger component which is caffeine reversible.

As shown in figure 1, the caffeine-insensitive tension loss occurs early during a fatigue run. Already after 1 min of stimulation application of caffeine only had a negligible potentiating effect and the same was true when the drug was applied just before the final, relatively rapid tension fall.

The main conclusion from the caffeine experiments on mouse fibers is similar to that for *Xenopus* fibers: a major part of the tension loss in fatigue appears to be due to inefficient activation of the contractile elements.

The mechanism behind the early (caffeine-insensitive) tension fall is not entirely clear. It would appear to be due to some direct inhibitory effect on cross-bridge action, most likely mediated by some metabolic change. Again, results from skinned fiber experiments would suggest H^+ and P_i as the most likely candidates. Preliminary experiments with pH indicators [Allen and Westerblad, personal commun.] indicate that under the present experimental conditions pH_i changes very little. This leaves an increase in P_i as the main cause, but it is not immediately clear why P_i should have a marked inhibitory effect during the first 10–15 tetani of a fatigue run and then be much less effective. A partial explanation might be that cross-bridge turnover rate decreases fairly rapidly in these fibers [18]

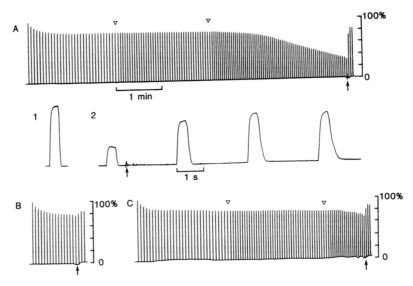

Fig. 1. The tension response to application of 15 m*M* caffeine (arrows) at various times during consecutive fatigue runs of a single mouse muscle fiber. All records obtained at the same gain; 100% refers to the amplitude of the first tetanus in each run. The time scale in *A* applies to *B* and *C* as well. The lower part of *A* shows selected records on an expanded time scale: 1 is the first tetanus and in 2 are shown the last tetanus before caffeine application and the three consecutive caffeine tetani. From Lännergren and Westerblad [7].

so that CrP hydrolysis and P_i production are slowed down. Also, an increasing contribution by oxidative metabolism to energy supply may reduce further CrP hydrolysis.

The mechanism behind the late (caffeine-sensitive) fall in tension also remains to be better understood. As in *Xenopus* fibers, it may in principle be due either to decreased Ca release, to decreased Ca sensitivity, or to both. Measurements of Ca transients during fatiguing stimulation using fura-2 have been started by Allen and Westerblad. Preliminary results suggest that the amplitude of the Ca transient decreases less in mouse than in *Xenopus* fibers during fatigue, which indicates that decreased Ca sensitivity is also an important cause. The factors responsible for such a change need to be identified. Obviously, it would be of great importance to have data for metabolite concentrations at various stages of fatiguing stimulation, but due to the small size of the preparation such data will be difficult to obtain.

Another important question is the coupling between metabolic capacity and the time to the final, fairly rapid fall in tension. Although we have not studied mouse fibers of different histochemical types, it is clear from many other studies that there is a strong correlation between endurance and oxidative metabolic capacity (e.g. [19]; experiments described above on different types of *Xenopus* fibers).

Slowing of Relaxation

Slowing of relaxation is a well-known attribute of muscle fatigue and has been documented in a large number of instances. It is also very clearly evident in the present preparations and can be analyzed in some detail. Relaxation in small, homogeneous preparations and in single fibers characteristically consists of two phases. There is an initial, nearly linear phase, which then yields to a more rapid, more or less exponential phase. It has been shown that during the first phase, relaxation is isometric with good homogeneity of the fiber whereas at the 'shoulder' segment movements start to appear and the pattern becomes more chaotic [20, 21]. Figure 2 shows changes in the relaxation phase during a fatigue run of a mouse fiber. The slowing sets in quite rapidly and then progresses to reach a maximum at the end of phase 2 (the phase of almost constant force generation). Not only does the rate decrease but the duration of the linear phase increases as well, which is likely to be due to segment movements becoming delayed [21].

The mechanism behind the slowing is not entirely clear. It is still an open question whether relaxation rate is primarily determined by SR Ca uptake rate or by cross-bridge detachment rate. In reality, the situation is probably complex with an interaction between the two processes.

To what, then, may the slowing be due? The most likely answer would seem to be that it is due to effects both on the SR and the cross-bridges. Shortening velocity decreases considerably during fatigue [7, 22], implying a lower cross-bridge turnover rate and hence also a slower detachment rate. However, it has also been shown by Ca_i measurements that the decline of the Ca transient is markedly slower in the fatigued state [12]. Thus, there might be an action at both sites.

An increase in H^+ concentration is presumably the major cause of slowed relaxation since increased P_i neither influences SR Ca pumps nor affects cross-bridge turnover rate [23]. Low pH has been shown both to decrease Ca sequestration by isolated SR vesicles [24] and to decrease shortening velocity [22]. Acidification of mouse fibers by exposure to 30% CO_2 causes a marked slowing of relaxation [25].

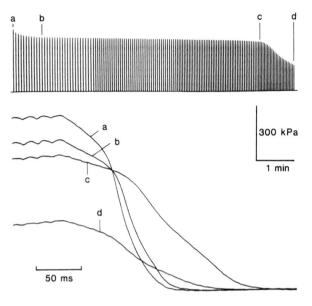

Fig. 2. Slowing of relaxation during fatiguing stimulation of a single mouse fiber. Each tetanus appears as a vertical line in the upper part. The lower part displays the relaxation of the tetani indicated above the fatigue curve. The tension bar refers to the upper part only. From Westerblad and Lännergren [25].

A substantial slowing of relaxation of fatigued *Xenopus* fibers, especially type 1 fibers, is also observed. The slowing is in fact more pronounced in *Xenopus* than in mouse fibers with a fall of relaxation rate in the former to about 25% of the rested value. A likely explanation for the difference would be, in line with the reasoning above, that during fatigue, pH_i falls more in *Xenopus* (type 1) than in mouse fibers.

Recovery after Fatiguing Stimulation
Generally speaking, the recovery process is of obvious importance from a functional point of view. Good recovery of isolated preparations also serves as a check that the preceding force decline represents genuine fatigue rather than a general running-down of the preparation.

The time course of force recovery depends on a number of factors such as type of fatiguing stimulation, fiber type, composition of external medium and temperature.

Force recovery after high-frequency stimulation is very rapid and probably reflects restitution of a normal ionic composition in the t tubules [4, 26].

Force recovery after intermittent tetanic stimulation is a considerably slower process than in the case above. Recovery is faster in type 3 fibers than in types 1 and 2. Especially in type 2 fibers, an initial rapid recovery is followed by a secondary tension fall (postcontractile depression, PCD [3]) before force production finally returns to the original level. During PCD, normal action potentials can be elicited but the amount of Ca released is extremely small [12]. The exact origin of PCD is not known, but evidence has been adduced that it reflects some reversible impairment of t tubule-SR communication [27].

In mouse fibers, force recovery has not yet been studied in detail. It occurs a good deal faster than in *Xenopus* fibers and so far no evidence of delayed force depression has been observed, using standard 70 Hz tetani as tests. However, if a lower stimulation frequency is used (e.g. 40 Hz), the force level attained during the initial part of recovery is lower than in the rested state. This resembles 'low-frequency fatigue' as first described by Edwards et al. [28] in human subjects, but is of shorter duration. Low-frequency fatigue has since been described in isolated mammalian muscles and in animal preparations. A possible connection with postcontractile depression in frog fibers is suggested by the finding that low-frequency fatigue can be counteracted by low doses of caffeine [29]

References

1   Lännergren J, Westerblad H: The temperature dependence of isometric contractions of single, intact fibres dissected from a mouse foot muscle. J Physiol 1987;390:285–293.
2   Lännergren J, Hoh JFY: Myosin isoenzymes in single muscle fibres of *Xenopus laevis*; Analysis of five different functional types. Proc R Soc 1984;B222:401–408.
3   Westerblad H, Lännergren J: Force and membrane potential during and after fatiguing, intermittent tetanic stimulation of single *Xenopus* muscle fibres. Acta Physiol Scand 1986;128:369–378.
4   Lännergren J, Westerblad H: Force and membrane potential during and after continuous high-frequency stimulation of single *Xenopus* muscle fibres. Acta Physiol Scand 1986;128:359–368.
5   Westerblad H, Lee JA, Lamb AG, Bolsover SR, Allen DG: Spatial gradients of intracellular calcium in skeletal muscle during fatigue. Pflügers Arch 1990;415:734–740.
6   Westerblad H, Lännergren J: The relation between force and intracellular pH in fatigued, single *Xenopus* muscle fibres. Acta Physiol Scand 1988;133:83–89.

7 Lännergren J, Westerblad H: Force decline due to fatigue and intracellular acidification in isolated fibres from mouse skeletal muscle. J Physiol 1991;434:307–322.

8 Cady EB, Jones DA, Lynn J, Newham DJ: Changes in force and intracellular metabolites during fatigue of human skeletal muscle. J Physiol 1989;418:311–325.

9 Völlestad NK, Sejersted OM, Bahr R, Woods JJ, Bigland-Ritchie BB: Motor drive and metabolic responses during repeated submaximal contractions in humans. J Appl Physiol 1988;64:1421–1427.

10 Cooke R, Franks K, Luciani GB, Pate E: The inhibition of rabbit skeletal muscle contraction by hydrogen ions and phosphate. J Physiol 1988;395:77–97.

11 Godt RE, Nosek TM: Changes of intracellular milieu with fatigue or hypoxia depress contraction of skinned rabbit skeletal and cardiac muscle. J Physiol 1989;412:155–180.

12 Allen DG, Lee JA, Westerblad H: Intracellular calcium and tension during fatigue in isolated single muscle fibres from Xenopus laevis. J Physiol 1989;415:433–458.

13 Lee JA, Westerblad H, Allen DG: Changes in tetanic and resting $[Ca^{2+}]_i$ during fatigue and recovery of single muscle fibres from Xenopus laevis. J Physiol 1991;433:307–326.

14 Rousseau EJ, LaDine J, Liu QY, Meissner G: Activation of Ca^{2+} release channel of skeletal muscle sarcoplasmic reticulum by caffeine and related compounds. Arch Biochem Biophys 1988;267:75–86.

15 Wendt IR, Stephenson DG: Effects of caffeine on Ca-activated force production in skinned cardiac and skeletal muscle fibres of the rat. Pflügers Arch 1983;398:210–216.

16 Fryer MW, Neering IR: Actions of caffeine on fast- and slow-twitch muscles of the rat. J Physiol 1989;416:435–454.

17 Lännergren J, Westerblad H: Maximum tension and force-velocity properties of fatigued, single Xenopus muscle fibres studied by caffeine and high K^+. J Physiol 1989;409:473–490.

18 Crow MT, Kushmerick MJ: Chemical energetics of slow- and fast-twitch muscles of the mouse. J Gen Physiol 1982;79:147–166.

19 Kugelberg E, Lindegren B: Transmission and contraction fatigue of rat motor units in relation to succinate dehydrogenase activity of motor unit fibres. J Physiol 1979;288:285–300.

20 Edman KAP, Flitney EW: Laser diffraction studies of sarcomere dynamics during 'isometric' relaxation in isolated muscle fibres of the frog. J Physiol 1982;329:1–20.

21 Curtin NA, Edman KAP: Effects of fatigue and reduced intracellular pH on segment dynamics in 'isometric' relaxation of frog muscle fibres. J Physiol 1989;413:159–174.

22 Edman KAP, Mattiazzi AR: Effects of fatigue and altered pH on isometric force and velocity of shortening at zero load in frog muscle fibres. J Muscle Res Cell Motil 1981;2:321–334.

23 Cooke R, Pate E: The effects of ADP and phosphate on the contraction of muscle fibres. Biophys J 1985;48:789–798.

24 MacLennan DH: Purification and properties of an adenosine triphosphatase from sarcoplasmic reticulum. J Biol Chem 170;245:4508–4518.

25 Westerblad H, Lännergren J: Slowing of relaxation during fatigue in single mouse muscle fibres. J Physiol 1991;434:323–336.

26   Jones DA: Muscle fatigue due to changes beyond the neuromuscular junction; in
     Porter R, Whelan J (eds): Human Muscle Fatigue: Physiological Mechanisms. Ciba
     Foundation Symposium 82. London, Pitman Medical, 1981, pp 178–192.
27   Westerblad H, Lännergren H: Recovery of fatigued *Xenopus* muscle fibres is markedly
     affected by the extracellular tonicity. J Muscle Res Cell Motil 1990;11:147–153.
28   Edwards RHT, Hill DK, Jones D, Merton PA: Fatigue of long duration in human
     skeletal muscle after exercise. J Physiol 1977;272:769–778.
29   Jones DA, Howell S, Roussos C, Edwards RHT: Low-frequency fatigue in isolated
     skeletal muscles and the effect of methylxanthines. Clin Sci 1982;63:161–167.

Dr. Jan Lännergren, Department of Physiology II, Karolinska Institutet,
S–104 01 Stockholm (Sweden)

Marconnet P, Komi PV, Saltin B, Sejersted OM (eds): Muscle Fatigue Mechanisms in
Exercise and Training. Med Sport Sci. Basel, Karger, 1992, vol 34, pp 54–68

Metabolic Aspects of Fatigue in Human Skeletal Muscle

Kent Sahlin

Department of Clinical Physiology, Karolinska Institute,
Huddinge University Hospital, Huddinge, Sweden

Introduction

Muscle fatigue is of central importance in exercise physiology and has
been the subject of numerous investigations. Our knowledge of the mechanism of the contraction process and its limits has been extended considerably
during recent years. However, the mechanism(s) and the cause(s) of fatigue is
(are) still not fully understood and indicate the complexity of the subject.
Muscle fatigue may be defined as a transient decrease in the capacity to
perform work due to prior physical activity [Asmussen, 1979] or as a failure
to maintain the required force or exercise intensity [Edwards, 1983]. Since
the latter definition is more suitable as an operational unit, it will be used in
the present paper.

Voluntary contraction is a complex series of events, and fatigue can
occur at various sites in the chain from the brain to force generation. A
distinction is usually made between central fatigue where the impairment is
located in the central nervous system and peripheral fatigue where the
impairment is located in the peripheral nerve or contracting muscle. Due to
the different feedback signal systems, a clear separation between central and
peripheral fatigue cannot always be made. Especially during prolonged
exercise the point of fatigue is less clear and more influenced by psychological
factors than during short-term intensive exercise. Although central fatigue is
an important factor for performance during many conditions considerable
evidence is available that in well-motivated subjects, a substantial component of fatigue can be located to the muscle [Bigland-Ritchie et al., 1984].
Central fatigue may also to some extent, bear a relation to changes in the

periphery mediated to the CNS by neural or chemical (plasma levels of glucose, NH3, K^+, H^+, amino acids) feedback systems.

Since muscle contraction requires a continuous supply of energy and since the energetic processes are limited in their rate and capacity, it seems obvious to focus on metabolic changes as a cause of fatigue. The purpose of this paper is to discuss the importance of metabolic factors as a cause of fatigue during voluntary exercise in humans. Other aspects of muscular fatigue have been discussed in this volume and in other reviews [Edwards 1981; Fittz and Metzger, 1988; Green, 1990; Sjøgaard, 1990; Vøllestad and Sejersted, 1988].

Exercise Endurance and Limitations of the Energetic Processes

The breakdown of ATP to ADP and the rephosphorylation of ADP back to ATP constitutes the ATP-ADP cycle by which the energy-consuming processes are coupled to the energy-yielding processes. The processes by which ATP is regenerated are characterized by the maximal rate of ATP generation and the amount of ATP that can be produced (fig. 1). These limitations of the metabolic processes will set an upper bound for the energy production and thus of the intensity and the duration at which exercise can be performed. The intensity of the exercise and the associated rate of energy demand will be an important determinant of the selection of fuel and, thus, the relation between aerobic/anaerobic energy utilization and CHO/FFA oxidation. Other factors of importance are the availability of oxygen and fuels, the relative capacity of the different energetic processes, and hormonal changes.

During sustained exercise the major part of energy is produced through the oxidative processes and it is therefore evident that the aerobic capacity will be an important determinant of endurance at a certain absolute work load. The difference between subjects in endurance can be diminished by expressing the work load in relation to the maximal oxygen uptake (%$\dot{V}_{O_2}$ max). The relation between endurance and exercise intensity shown in figure 2 has been constructed from results obtained from studies where subjects were physically active but not endurance trained at the top level. There is a rightward shift in the curve in endurance-trained subjects and a leftward shift in sedentary subjects.

The relation between endurance and intensity has a strong curvilinear appearance and is probably determined to a large extent by metabolic

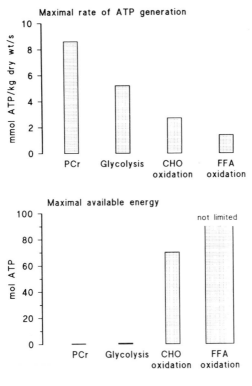

Fig. 1. Characteristics of the energetic processes in human skeletal muscle. The figure has been constructed from the data presented in tables 1 and 2 of Sahlin [1986]. The maximal rate of aerobic ATP production corresponds to a subject with a $\dot{V}_{O_2}$ max of 4.0 liters/min and a working muscle mass of 20 kg. The available energy was estimated from 20 kg muscle with a glycogen content (glucosyl units) of 70 mmol/kg wet eight and 500 mmol of liver glycogen. Available energy from glycolysis corresponds to a condition where cellular accumulation of lactate limits further glycolysis.

factors. At about 50% $\dot{V}_{O_2}$ max, the curve has an asymptote and endurance approaches infinity. This is probably related to the fact that combustion of fat can provide the whole energy requirement at work loads below about 50% $\dot{V}_{O_2}$ max [Newsholme and Leech, 1983] and that the storage of fat is abundant. The point of fatigue is unclear at these low work loads and is, to a large extent, determined by factors influencing the CNS such as pain, discomfort, dehydration and increased temperature.

At intensities between 60 and 90% of $\dot{V}_{O_2}$ max, fatigue is associated with depleted stores of muscle glycogen [Saltin and Karlsson, 1972] but since

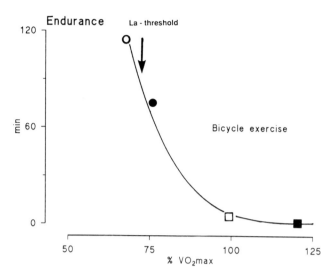

Fig. 2. Endurance during cycling in relation to the relative intensity. The curve is shifted to the right in endurance-trained subjects and to the left in sedentary subjects. The symbols correspond to the average values in Björkman et al. [1984] [○]; Sahlin et al. (1990) [●]; Sahlin et al. [1989] [□]; Medbø and Tabata (1989) [■].

muscle lactate is only moderately elevated, muscle pH is close to the pre-exercise level (table 1). Endurance at these intensities is closely related to the pre-exercise muscle level of glycogen [Hermansen et al., 1967] and to the rate of glycogen depletion. Blood lactate reflects the formation of lactate in the working muscle and is therefore related to the rate of glycogen degradation. The running speed at the lactate threshold has been shown to be closely related to the performance during marathon running [Sjödin and Svedenhag, 1985]. The link between lactate and fatigue is, in this case, not related to acidosis but most likely to the rate of glycogen depletion.

At higher intensities ($>90\%$ $\dot{V}_{O_2}$ max) an increasing part of the energy requirement is covered by anaerobic processes. This results in accumulation of metabolic end products and a perturbation of the cellular homeostasis (table 1). At these high intensities fatigue is characterized by a marked depletion of high-energy phosphates by a decrease in pH due to lactate accumulation, but in maintained glycogen levels.

Table 1. Exercise-induced changes in muscle metabolites (mmol/kg dry weight), pH and glycogen (% of initial value)

	Rest	Isometric contraction to fatigue[a]	Cycling to fatigue[b,c]	Cycling to fatigue[d]
Intensity	–	66% MVC	100% $\dot{V}_{O_2}$ max	75% $\dot{V}_{O_2}$ max
Duration, min	–	0.8	6	75
Lactate	3	91	113	14
Muscle pH	7.1	6.6	6.6	7.1[e]
ATP	25	21	20	23
ADP	3	3.5	3.8	3.5
IMP	0	3[f]	3.5	2
Phosphocreatine	87	7	17	29
P_i	38[g]	108[h]	88[g]	100[h]
$H_2PO_4^-$	13	68	55	34
Glycogen	100	85	81[i]	11

[a]Sahlin et al. [1975].   [c]Sahlin et al. [1989].
[b]Sahlin et al. [1976].   [d]Sahlin et al. [1990].
[e]Estimated from changes in PCr and lactate.
[f]Sahlin and Ren [1989].
[g]Sahlin et al. [1978].
[h]Calculated from changes in PCr, ATP, ADP, glucose-6-P.
[i]Estimated from changes in lactate and glucose-6-P.

Hypothesis: Insufficient Energy Supply Impairs Force Generation

A classical theory of fatigue is that this corresponds to a situation where the ATP consumption cannot be met by an equal rate of ATP formation and the contraction process is impaired by energetic deficiency. However, the cellular ATP content remains practically unchanged during exercise to fatigue (maximum decrease about 10–30%) and the theory is, therefore, generally rejected. This line of argument may, however, be too simplistic an approach to the problem.

The cellular content of ADP is about 10 times lower than ATP and since the major part of ADP is protein bound, the free ADP which is the metabolic active form is even lower. Although it seems unlikely that the small decline in the concentration of ATP can limit the force generation, a small decline in ATP will cause a major relative increase in ADP and inorganic phosphate (P_i). Energetic deficiency will therefore, despite almost unchanged ATP levels, be manifested by large relative increases in ADP and P_i (table 1), which are potential fatiguing agents.

The rate of ATP utilization during high intensity exercise (>200 mmol/ kg dry weight/min) increases more than 100 times from that at rest and corresponds to a turnover of the whole store of free ADP (and total ADP) in 5 ms (1 s). The true concentrations of ADP and AMP in a contracting muscle at the cellular sites of ATP hydrolysis are therefore uncertain and cannot be assessed by currently available techniques. Under conditions of a high ATP turnover and a low cellular PCr concentration, it is likely that both temporal and spatial concentration gradients of ADP will exist within the cell [Funk et al., 1989].

It has previously been suggested [Sahlin, 1986] that transient increases in ADP and AMP occur in a contracting muscle at the site of ATP hydrolysis. During conditions of energetic deficiency, the maximal rate of ADP rephosphorylation will decrease and the transient increase in ADP (and AMP) are expected to be extended in time and amplitude. It was further suggested [Sahlin, 1986] that these local transient increases in ADP are related to a failure of the contraction process. A decrease in the maximal capacity to rephosphorylate ADP may not necessarily be manifested by measurable changes in ATP or ADP since the increases in ADP are likely to be too small and short-lasting to change the total ADP content averaged over time and space.

An alternative approach would be to estimate the maximal rate of ADP rephosphorylation. A low PCr level would severely limit the maximal rate of ADP rephosphorylation since the creatine kinase reaction has a high power (fig. 1). It is notable that the decline in force during a sustained static contraction and the restoration of force during the recovery period are closely related to the PCr level [Sahlin and Ren, 1989]. Similarly, the frequently observed association between acidosis and fatigue may be due to an acidotic impairment of ADP rephosphorylation (see below). Potential links between metabolic changes and decline in force are shown in figure 3.

Significance of AMP Deamination in Fatigue

Increases in AMP and ADP are potent stimulators of AMP deaminase [Wheeler and Lowenstein, 1979], which catalyzes the deamination of AMP to inosine monophosphate (IMP) and amonia (NH_3). Since further metabolism of IMP is a rather slow process, the muscle content of IMP may reflect the AMP and ADP transients and the existence of energetic deficiency.

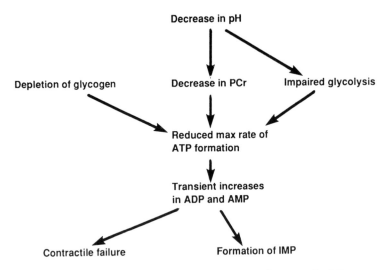

Fig. 3. Potential links between metabolic changes and contractile failure.

The content of adenine nucleotides, IMP, and NH_3 have been measured in needle biopsy specimens taken from the human quadriceps femoris muscle at rest and after different types of exercise [Sahlin and Broberg, 1990]. During low-intensity exercise ($<50\%$ $\dot{V}_{O_2}$ max) there is no measurable increase in IMP or NH_3, which demonstrates that the rate of AMP deamination is low. At high intensities both IMP and NH_3 accumulate in the working muscle and the increases are stoichiometrical to the decrease in TAN (total adenine nucleotides = ATP + ADP + AMP). The formation of IMP seems to be related to metabolic stress since the level of IMP is related to lactate and to the decrease in PCr [Sahlin et al., 1989]. About 15% of the AN pool is degraded to IMP after exercise to fatigue at 100% of $\dot{V}_{O_2}$ max [Sahlin and Broberg, 1990].

Prolonged exercise to fatigue leading to low intramuscular levels of glycogen is also associated with increased deamination of AMP [Norman et al., 1987; Broberg and Sahlin, 1989; Sahlin et al., 1990] and the formation of IMP is inversely related to the glycogen level. Furthermore, exercise with low initial glycogen levels results in a rapid development of fatigue and in a more pronounced formation of IMP and NH_3 than during exercise with normal pre-exercise glycogen levels [Broberg and Sahlin, 1989]. If carbohydrates are supplied at the end of exercise, the duration can be prolonged and

Table 2. Conditions which impair work capacity and result in an enhanced catabolism of adenine nucleotides during exercise.

Impaired glycogen utilization	
Glycogen depletion	Norman et al. [1987]; Broberg and Sahlin [1989]
McArdle's disease	Mineo et al. [1985]; Sahlin et al. [1990]
Impaired glycolysis	
PFK-deficient patients	Kono et al. [1986]
Acidosis	Dudley and Terjung [1985]; Greenhaff et al. [submitted]
Hypoxia	Sahlin and Katz [1989]
β-Adrenoceptor blockade	Broberg et al. [1988]

the IMP formation can be attenuated [Spencer et al., 1991]. These studies indicate that decreased levels of carbohydrates result in an energetic deficiency at the adenine nucleotide level and that the mechanism of fatigue may be similar as during high-intensity exercise.

An enhanced formation of IMP and NH_3 has also been observed during other conditions of impaired exercise capacity (table 2). Thus, in McArdle patients, who have a reduced work capacity, the increase in muscle IMP and plasma NH_3 were excessive when compared with control subjects exercising at the same absolute workload [Sahlin et al., 1990]. During β-adrenoceptor blockade (βB) exercise capacity decreases due to both a reduced O_2 supply and peripheral metabolic factors [Tesch, 1985]. Exercise with βB resulted in an almost 2-fold higher increase in IMP than during exercise without βB for the same duration [Broberg et al., 1988]. Furthermore, exercise with a reduced arterial O_2 tension, which impairs the aerobic work capacity, results in increased muscle IMP levels as compared with control conditions [Sahlin and Katz, 1989]. Similarly, acute exposure to high altitude resulted in the same increase in plasma NH_3 during exercise despite a lower absolute work load than during control conditions [Young et al., 1987].

A decreased capacity to rephosphorylate ADP in combination with a high rate of ATP turnover appears to be a common metabolic denominator for conditions resulting in increased deamination of AMP and impaired work capacity. This demonstrates that muscle fatigue during many conditions is associated with an enhanced catabolism of the adenine nucleotide pool and thus suggests that energetic deficiency is likely to be a major cause of fatigue.

Energetic deficiency could be linked to an impaired muscle function either through accumulation of ADP or through increases in P_i, which is formed when PCr decreases. From studies on skinned muscle fibers it has been shown that increases of P_i to the physiological level impairs both the isometric tension and the force-velocity curve [Cooke and Pate, 1990]. Some studies in man have demonstrated that the concentration of the diproteinated form of P_i ($H_2PO_4^-$) is closely related to the decline in force [Wilson et al., 1988] but other studies have shown a dissociation between these two factors [Cady et al., 1989; Hultman et al., 1990].

Acidosis and Failure of Force Generation

Accumulation of lactic acid is, in many cases, related to a decline in force generation, and a decrease in muscle pH is often considered as a main fatiguing agent. Acidosis could impair the contraction process through different mechanisms. One possibility is that increased hydrogen ion concentration interferes with the energy supply, which secondarily affects one or several steps in the contraction process. Another possibility is that increased H^+ ion concentration interferes directly with the contractile machinery or the excitation-contraction coupling. This latter hypothesis is supported by findings from in vitro studies with muscle fibers where the cell membrane has been removed (skinned fibers) and the intracellular chemical composition altered [Cooke and Pate, 1990; and others].

The time courses of the recovery in force and metabolites have recently been studied in man after a fatiguing contraction. Maximal force reverted to the pre-contraction value with a half-time of about 15 s and after 2 min of recovery force was not statistically different from that before contraction. In contrast, recovery in lactate and muscle pH occurred with a slower time course and after 2 min recovery calculated muscle pH was similar to the immediate postcontraction value [Sahlin and Ren, 1989]. These data in humans demonstrate that near maximal contraction force can be attained during acidotic conditions. Consequently, the large depression of force at low pH observed in vitro with the skinned fiber preparation does not apply to humans in vivo. The link between H^+ accumulation and muscle fatigue is therefore probably not a simple cause and effect but rather indirect. The inhibitory effect of acidosis on glycolysis and the involvement of H^+ in the creatine kinase equilibrium [Sahlin, 1986] may explain the connection between acidosis and impairment of force generation which has been observed in vivo.

Table 3. Effect of altered contraction characteristics on performance during static and dynamic exercise

	Static	Dynamic
Reduced isometric force	negative	negative
Slowing of relaxation	positive	negative
Reduced shortening velocity	no effect	negative

Slowing of Relaxation and Muscle Metabolic Changes

It has been shown [Viitasalo and Komi, 1981] that prolonged static contraction results not only in a reduced maximal force but also in a reduced rate of relaxation and a reduced rate of force development. Much less is known about how these contractile parameters changes after dynamic exercise. Some data indicate that prolonged dynamic exercise also results in a decreased force, relaxation rate and rate of force development [Viitassalo et al., 1982].

A decrease in maximal force will impair both static and dynamic exercise (table 3). Muscle fatigue is therefore generally considered as an effect of reduced force generation and sometimes the terms are used synonymously. However, reduced rates of relaxation and force development will also have a negative influence on dynamic exercise. In contrast, slowing of relaxation will enhance the endurance of a contraction, since a certain force can then be maintained at a lower cross-bridge turnover and thus at a lower energy demand. The rates of force development and relaxation are therefore important factors for the performance but will affect dynamic and static exercise differently.

The cause of the reduced relaxation rate is not known. Some studies have shown that the slowing of relaxation is closely related to acidosis [Sahlin et al., 1981; Edman and Mattiazzi, 1981] whereas other studies have demonstrated that changes in high-energy phosphates are more important [Edwards et al., 1975; Sjöholm et al., 1983]. In a recent study [Cady et al., 1989], it was shown that the relaxation rate decreased during static contraction also in a myophosphorylase-deficient patient, despite unchanged muscle pH. The change was, however, less than in control subjects and, furthermore, the relaxation rate was more rapidly reversed during the subsequent recovery period than in control subjects. The authors concluded that slowing of relaxation was caused both by a pH-dependent and a pH-independent factor

[Cady et al., 1989]. In a study of single fibers of frog muscle, changes in the cellular Ca concentration was measured with aequorin [Allen et al., 1989]. It was concluded that decrease in pH reduced the relaxation rate through a mechanism independent of Ca removal, but that the slower Ca removal in fatigued muscle suggested a reduced rate of Ca pumping by the SR system, possibly due to a decrease in the free-energy change of ATP hydrolysis [Dawson et al., 1980].

Energetic Deficiency and Fatigue: Potential Mechanisms

From the discussion above different lines of evidence suggest that energetic deficiency is an important factor in fatigue during both short-term and prolonged exercise. Several steps in the contractile process are known to be energy dependent: maintenance of the Na-K gradient over sarcolemma; re-uptake of Ca^{2+} by the sarcoplasmatic reticulum (SR); cross-bridge cycling; coupling between t tubular depolarization and Ca release from SR [suggested by Donaldson, 1990].

Experiments with the skinned fiber preparation, which provides information about the cross-bridge interaction, have shown that the force and the force-velocity are almost unaffected by changes in ADP and ATP within the physiological range, whereas changes in P_i and H^+ have a major influence [Cooke and Pate, 1990].

An impaired function of the re-uptake of Ca^{2+} by SR would result in a slowing of relaxation and some data suggest that this step is impaired by energetic deficiency (see above). Further studies are required to elucidate the exact mechanism.

Maintenance of the membrane potential and the trans-membrane Na-K gradient requires an expenditure of energy. The free-energy change of ATP hydrolysis will determine the maximal gradient that can be maintained and some studies suggest that the decrease in the free-energy change that occurs in a fatigued muscle is sufficient to affect the Na-K gradient [Fiolet et al., 1984; Blum et al., 1988]. The increase in intracellular Na and decrease in K, which is observed in muscle at fatigue [Sahlin et al., 1977; Sjøgaard et al., 1985], may, therefore, be a consequence of energetic deficiency. The resulting decrease in membrane potential may impair action potential transmission [Hodgkin and Huxley, 1952]. Alternatively, it has been suggested [Sjøgaard, 1990] that a decrease in cellular ATP could activate ATP sensitive K channels and increase the K conductance, thereby impairing the activation of the membrane [Sjøgaard, 1990].

Concluding Remarks

The present paper has focused on the role of metabolic factors in fatigue. Many studies demonstrate a close relation between the metabolic capacity to produce ATP and exercise performance. Energetic deficiency is therefore likely to play a major role in the etiology of muscle fatigue. At fatigue, characteristic metabolic changes are generally observed – the pattern being different after short-term exercise (La accumulation and PCr depletion) than after prolonged exercise at moderate intensity (glycogen depletion). A common metabolic denominator at fatigue during these conditions is a decreased capacity to generate ATP coupled to a high ATP turnover and is expressed in the cell as an increased catabolism of the adenine nucleotide pool. Previous attempts to find a sole factor responsible for fatigue during various conditions have been fruitless. In the present paper transient increases in ADP is suggested as a link between energetic deficiency and fatigue. Although the increased rate of AMP deamination during conditions of high rates of energy turnover and energetic deficiency may be taken as a support for the hypothesis of exercise-induced ADP-AMP transients, the theory remains speculative, but may serve the purpose of stimulating further research.

The present report demonstrates that metabolic factors play an important role in muscle fatigue in vivo but there is no doubt that conditions exist where fatigue cannot be explained by an altered metabolic state [Jones, 1981; Edwards, 1981; Vøllestad and Sejersted, 1988; Hultman et al., 1990; Green, 1990]. This is not surprising considering the complexity of the contraction process and the potential variability of the contraction, which may alter the limiting site and thereby also the mechanism of fatigue.

References

Allen DG, Lee JA, Westerblad H: Intracellular calcium and tension during fatigue in isolated single muscle fibres from xenopus laevis. J Physiol 1989;415:433–458.

Asmussen E: Muscle fatigue. Med Sci Sports 1979;11:313–321.

Bigland-Ritchie B, Woods JJ: Changes in muscle contractile properties and neural control during human muscular fatigue. Muscle Nerve 1984;7:691–699.

Björkman O, Sahlin K, Hagenfeldt L, Wahren J: Influence of glucose and fructose ingestion on the capacity for long-term exercise in well-trained men. Clin Physiol 1984;4:483–494.

Blum H, Schnall MD, Chance B, Buzby GP: Intracellular sodium flux and high-energy phosphorus metabolites in ischemic skeletal muscle. Am J Physiol 1988;255: C377–C384.

Broberg S, Katz A, Sahlin K: Propranolol enhances adenine nucleotide degradation in human muscle during exercise. J Appl Physiol 1988;65:2478–2483.

Broberg S, Sahlin K: Adenine nucleotide degradation in human skeletal muscle during prolonged exercise. J Appl Physiol 1989;67:116–122.

Cady EB, Elshore H, Jones DA, Moll A: The metabolic causes of slow relaxation in fatigued human skeletal muscle. J Physiol 1989;418:327–337.

Cady EB, Jones DA, Lynn J, Newham DJ: Changes in force and intracellular metabolites during fatigue of human skeletal muscle. J Physiol 1989;418:311–325.

Cooke R, Pate E: The inhibition of muscle contraction by the products of ATP hydrolysis; in Taylor AW, Gollnick PD, Green HJ, Noble EG, Sutton JR (eds): Biochemistry of Exercise VII, pp 59–72. Champaign, Human Kinetics, 1990.

Dawson MJ, Gadian DG, Wilkie DR: Mechanical relaxation rate and metabolism studied in fatiguing muscle by phosphorous nuclear magnetic resonance. J Physiol 1980;299:465–484.

Donaldson SKB: Fatigue of sarcoplasmic reticulum. Failure of excitation-contraction coupling in skeletal muscle, in Taylor AW, Gollnick PD, Green HJ, Ianuzzo CD, Noble EG, Metivier G, Sutton JR (eds): Biochemistry of Exercise VII, pp 49–57. Champaign, Human Kinetics, 1990.

Dudley GA, Terjung RL: Influence of acidosis on AMP deaminase activity in contracting fast-twitch muscle. Am J Physiol 1985;248:C43–C50.

Edman KAP, Mattiazzi AR: Effect of fatigue and altered pH on isometric force and velocity of shortening at zero load in frog muscle fibers. J Musc Res Cell Motil 1981;2:321–334.

Edwards RHT, Hill DK, Jones DA: Metabolic changes associated with the slowing of relaxation in fatigued mouse muscle. J Physiol 1975;251:287–301.

Edwards RHT: Human muscle function and fatigue; in Porter R, Whelan J (eds): Human Muscle Fatigue: Physiological Mechanisms, pp 1–18. London, Pitman, 1981.

Edwards RHT: Biochemical bases of fatigue in exercise performance: Catastrophe theory of muscular fatigue; in Vogel HG, Knuttgen HG, Poortmans J (eds): Biochemistry of Exercise, Int Ser Sports Sci, pp 3–28. Champaign, Human Kinetics, 1983.

Fiolet JWT, Baartscheer A, Schumacher CA, Coronel R, ter Welle HF: The change of the free energy of ATP hydrolysis during global ischemia and anoxia in the rat heart. J Mol Cell Cardiol 1984;16:1023–1036.

Fitts RH, Metzger JM: Mechanisms of muscular fatigue; in Poortmans JR (ed): Principles of Exercise Biochemistry, pp 212–229. Basel, Karger, 1988.

Funk C, Clark A, Connett RJ: How phosphocreatine buffers cyclic changes in ATP demand in working muscle. Adv Exp Med Biol 1989;248:687–692.

Green HJ: Manifestations and sites of neuromuscular fatigue; in Taylor AW, Gollnick PD, Green JH, Ianuzzo CD, Noble EG, Metivier G, Sutton JR (eds): Biochemistry of Exercise VII, pp 13–35. Champaign, Human Kinetics, 1990.

Greenhaff PL, Harris RC, Snow DH, Sewell DA, Dunnett M: The influence of metabolic alkalosis upon exercise metabolism in the thoroughbred horse. Europ J Appl Physiol (in press).

Hermansen L, Hultman E, Saltin B: Muscle glycogen during prolonged severe exercise. Acta Physiol Scand 1967;71:129–139.

Hodgkin AL, Huxley AF: The dual effect of membrane potential on sodium conductance in the giant axon of Liligo. J Physiol (Lond) 1952;116:497–506.

Hultman E, Bergström M, Spriet LL, Söderlund K: Energy metabolism and fatigue; in Taylor AW, Gollnick PD, Green JH, Ianuzzo CD, Noble EG, Metivier G, Sutton JR (eds): Biochemistry of Exercise VII, pp 73–92. Champaign, Human Kinetics, 1990.

Jones DA: Muscle fatigue due to changes beyond the neuromuscular junction; in Porter R, Whelan J (eds): Human Muscle Fatigue: Physiological Mechanisms. Ciba Found Symp 82, pp 178–192. London, Pitman, 1981.

Kono N, Mineo I, Shimizu T, Hara N, Yamada Y, Nonaka K, Tarni S: Increased plasma uric acid after exercise in muscle phosphofructokinase deficiency. Neurology 1986; 36:106–108.

Medbø JI, Tabata I: Relative importance of aerobic and anaerobic energy release during short-lasting exhausting bicycle exercise. J Appl Physiol 1989;67:1881–1886.

Mineo I, Kono N, Shimizu T, Hara N, Yamada Y, Sumi S, Nonaka K, Tarui S: Excess purine degradation in exercising muscles of patients with glycogen storage disease types V and VII. J Clin Invest 1985;76:556–560.

Newsholme EA, Leech AR: Biochemistry for the Medical Sciences. New York, Wiley, 1984.

Norman B, Sollevi A, Kaijser L, Jansson E: ATP breakdown products in human skeletal muscle during prolonged exercise to exhaustion. Clin Physiol 1987;7:503–510.

Sahlin K: Metabolic changes limiting muscle performance; in Saltin B (ed): Biochemistry of Exercise VI, pp 323–343. Champaign, Human Kinetics, 1986.

Sahlin K, Alvestrand A, Bergström J, Hultman E: Intracellular pH and bicarbonate concentration as determined in biopsy samples from the quadriceps muscle of man at rest. Clin Sci Mol Med 1977;53:459–466.

Sahlin K, Areskog NH, Haller PG, Henriksson KG, Jorfeldt L, Lewis SF: Impaired oxidative metabolism increases adenine nucleotide breakdown in McArdle's disease. J Appl Physiol 1990;69:1231–1235.

Sahlin K, Broberg S: Adenine nucleotide depletion in human muscle during exercise: Causality and significance of AMP deamination. Int J Sport Med 1990;II:S62–S67.

Sahlin K, Broberg S, Ren JM: Formation of IMP in human skeletal muscle during incremental dynamic exercise. Acta Physiol Scand 1989;136:193–198.

Sahlin K, Edström L, Sjöholm H, Hultman E: Effects of lactic acid accumulation and ATP decrease on muscle tension and relaxation. Am J Physiol 1981;240:C121–C126.

Sahlin K, Harris KC, Hultman E: Creatine kinase equilibrium and lactate content compared with muscle pH in tissue samples obtained after isometric exercise. Biochem J 1975;152:173–180.

Sahlin K, Harris RC, Nylind B, Hultman E: Lactate content and pH in muscle samples obtained after dynamic exercise. Pflügers Arch 1976;367:143–149.

Sahlin K, Katz A: Hypoxemia increases the accumulation of IMP in human skeletal muscle during submaximal exercise. Acta Physiol Scand 1989;136:199–203.

Sahlin K, Katz A, Broberg S: Tricarboxylic acid cycle intermediates in human muscle during prolonged exercise. Am J Physiol 1990;259:C834–C841.

Sahlin K, Palmskog G, Hultman E: Adenine nucleotide and IMP contents of the quadriceps muscle in man after exercise. Pflügers Arch 1978;374:193–198.

Sahlin K, Ren JM: Relationship of contraction capacity to metabolic changes during recovery from a fatiguing contraction. J Appl Physiol 1989;67:648–654.

Saltin B, Karlsson J: Muscle glycogen utilization during work of different intensities; in Pernow B, Saltin B (eds): Muscle Metabolism during Exercise, pp 289–299. New York, Plenum Press, 1972.

Sjödin B, Svedenhag J: Applied physiology of marathon running. Sports Med 1985;2:83–99.

Sjøgaard G: Exercise-induced muscle fatigue: The significance of potassium. Acta Physiol Scand 1990;140(suppl 593).

Sjøgaard G, Adams RP, Saltin B: Water and ions shifts in skeletal muscle of humans with intense dynamic knee extension. Am J Physiol 1985;248:R190–R196.

Sjöholm H, Sahlin K, Edström L, Hultman E: Quantitative estimation of anaerobic and oxidative energy metabolism and contraction characteristics in intact human skeletal muscle in response to electrical stimulation. Clin Physiol 1983;3:227–239.

Spencer MK, Yan Z, Katz A: Carbohydrate supplementation attenuates IMP accumulation in human muscle during prolonged exercise. Am J Physiol 1991;261: C71–C76.

Tesch PA: Exercise performance and beta-blockade. Review article. Sports Med 1985;2:389–412.

Wheeler T J. Lowenstein JM: Adenylate deaminase from rat muscle. Regulation by purine nucleotides and orthophosphate in the presence of 150 mM KCl. J Biol Chem 1979;254:8994–8999.

Viitasalo JT, Komi PV: Effects of fatigue on isometric force- and relaxation-time characteristics in human muscle. Acta Physiol Scand 1981;111:87–95.

Viitasalo JT, Komi PV, Jacobs I, Karlsson J: Effects of prolonged cross-country skiing on neuromuscular performance; in Komi PV (ed): Exercise and Sport Biology, vol 12. Int Series on Sport Sciences. Champaign, Human Kinetics, 1982.

Vøllestad NK, Sejersted OM: Biochemical correlates of fatigue. Eur J Appl Physiol 1988;57:336–347.

Wilson JR, McCully KK, Mancini DM, Boden B, Chance B: Relationship of muscular fatigue to pH and diprotenated P$_i$ in humans: A 31P-NMR study. J Appl Physiol 1988;64:2333–2339.

Young PM, Rock PB, Fulco CS, Trad LA, Forte VA, Cymerman A: Altitude acclimatization attenuates plasma ammonia during submaximal exercise. J Appl Physiol 1987; 63:758–764.

Kent Sahlin, PhD, Department of Clinical Physiology, Karolinska Institute, Huddinge University Hospital, S-141 86 Huddinge (Sweden)

Marconnet P, Komi PV, Saltin B, Sejersted OM (eds): Muscle Fatigue Mechanisms in Exercise and Training. Med Sport Sci. Basel, Karger, 1992, vol 34, pp 69–86

Role of Amino Acids and Ammonia in Mechanisms of Fatigue

Anton J.M. Wagenmakers

Nutrition Research Centre, Department of Human Biology, University of Limburg, Maastricht, The Netherlands

Introduction

Prolonged physical activity, such as that occurring in the marathon or cycling races leads to fatigue. Fatigue is defined physiologically as the inability to maintain power output [1]; to the endurance athlete it means an increased sense of effort, sometimes even pain and discomfort, and the need to reduce pace. A voluntary muscle contraction is the final step in a command chain that extends from the higher centres of the central nervous system to the actin and myosin filaments in the muscle itself; it involves electrical, biochemical and mechanical events. Consequently, many possible sites and mechanisms may be implicated in the fatigue process [1].

In this paper we will reflect on the biochemical causes of fatigue in endurance exercise especially with respect to the maximal aerobic oxidation rate of carbohydrates and fat. During long-distance running both carbohydrates and fatty acids are used to meet the energy requirements. However, as the glycogen stores in the muscle become depleted the rate of glycolysis and hence that of glucose oxidation will decrease progressively. It has been argued that the maximal rate at which fatty acids can be used during aerobic exercise is limited to about 50% of the maximum aerobic rate of ATP production [2–4]. This means that when the glycogen stores are depleted in the muscles of the runner, the pace must fall by 50%. The rate of uptake of fatty acids has been proposed as the limiting factor for the rate of ATP formation from fatty acid oxidation. However, no experimental evidence has been presented to support this proposal. In this paper reflections are made on the interactions between the metabolism of carbohydrates, fatty acids, amino acids and ammonia during exhaustive endurance exercise leading to glycogen depletion. Attention has been given in the past to the possible role of amino acids, especially the branched-chain amino acids (BCAA; leucine, isoleucine and valine) as the 'third fuel' during exercise [5]. However, no attention has been paid to the possible role of the BCAA as regulators of the flux in the citric acid cycle and therefore as regulators of the maximal aerobic rate of fuel oxidation and ATP production. Investigations in patients with McArdle's disease (a rare glycogen breakdown defect) have led us to the belief that BCAA play a central role in the biochemical mechanisms of peripheral fatigue and by presenting the lines of evidence which have led us to that belief and going through some experimental evidence that this mechanism may also apply to healthy athletes, we hope to convince the participants of this meeting and readers of this paper that amino acids cannot be ignored if we ever want to understand the complete integrative mechanism of fatigue.

BCAA also have been implicated in central mechanisms of fatigue through their modulating effect on tryptophan entry into the brain and therefore on serotonin synthesis [4, 6, 7]. Exercise-induced hyperammonemia has been implicated in central mechanisms of fatigue too and in extremis even in loss of motor control, coherent thought and consciousness [8, 9]. Critical reflections on these issues will be presented in the final sections of this paper.

Clues for an Alternative Peripheral Fatigue Mechanism from McArdle's Disease

The basis for an alternative hypothesis on the biochemical causes of muscle fatigue has been derived from a number of investigations which we recently performed in patients with McArdle's disease [10]. Due to a defect in the muscle phosphorylase enzyme these patients cannot use muscle glycogen as an energy source during exercise. They therefore are an ideal model ('an experiment of nature') in which to learn about the metabolic adaptations which develop during endurance exercise leading to glycogen depletion and exhaustion. Two early observations which we made in these patients are at the basis of the alternative hypothesis and are described in the next paragraphs.

The branched-chain 2-oxo acid dehydrogenase complex (BC-complex) is the key enzyme in the degradative pathway of the BCAA in muscle and in the early 1980s its activity was shown to be regulated by a phosphorylation/ dephosphorylation cycle (similar to the regulation of the pyruvate dehydrogenase complex). During exercise the total amount of BC-complex activity in human muscle does not change, but the proportion of enzyme present in the active (= dephosphorylated) form increases [11]. A greater proportion of the BC-complex was active in muscle in McArdle's disease at rest and following exhaustive exercise than in healthy individuals [10, 11]. Activation in McArdle's disease, furthermore, occurred at a much lower power output than in normal subjects and may occur very rapidly (one patient was exercised to exhaustion by cycling for 6 min at 60 W followed by 1.5 min at 80 W; the percentage active form of the BC-complex increased from 17 to 53% [10]). Wahren et al. [12] found a larger uptake of BCAA by exercising leg muscles in a patient with McArdle's disease than in normal individuals at comparable exercise intensity. These data suggest that the metabolism of BCAA is accelerated during exercise in patients with McArdle's disease as compared with healthy individuals.

A second remarkable observation made in McArdle's disease during incremental exhaustive cycling exercise and exercise at constant power output were excessive (up to 500 μM) increases in the forearm venous plasma ammonia concentration [10]. Femoral arteriovenous difference studies showed that the leg muscles were the origin of the ammonia production. During exercise at 100% W max the leg release of ammonia in a patient with McArdle's disease was 750 µmol/min [10], while in healthy individuals exercising at 80 and 100% $\dot{V}_{O_2}$ max values of 46 [13] and 89 µmol/min [14] have been reported. Ammonia production during exercise in McArdle's disease greatly exceeded the reported breakdown of ATP to IMP and therefore most likely originates from the metabolism of amino acids [10]. Deamination of amino acids via the reactions of the purine nucleotide cycle and/or via glutamate dehydrogenase are possible pathways [10].

To understand the relationship between acceleration of BCAA metabolism, excessive muscle production of ammonia and a potential mechanism of fatigue, we have to understand in detail the biochemical mechanisms normally used for disposal of the amino group of the BCAA and for disposal of ammonia. The BCAA are the main group of amino acids oxidized in muscle and therefore are the main amino group donors in muscle [5, 10]. In the BCAA aminotransferase reaction the amino group is donated to 2-oxoglutarate to form glutamate (fig. 1). In the reaction catalyzed by glutamine synthase glutamate subsequently reacts with ammonia, produced by the AMP-IMP reaction or deamination of amino acids, to form glutamine, the main

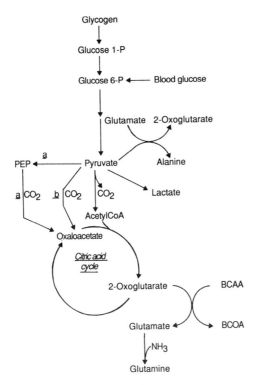

Fig. 1. Alternative biochemical mechanism of fatigue. BCAA metabolism is accelerated during exercise. The BCAA aminotransferase reaction is draining the carbon flux in
the citric acid cycle by using 2-oxoglutarate as an amino group acceptor. The draining
effect of the BCAA aminotransferase reaction is normally counteracted by the anaplerotic conversion of glycogen and glucose to citric acid cycle intermediates via carboxylation of pyruvate (route *b*) and/or reversal of the phosphoenolpyruvate carboxykinase
reaction (route *a*). Muscle glycogen therefore may also act as a precursor for the synthesis
of the carbon skeletons of glutamate and glutamine. In patients with glycogen breakdown
defects and in exercise leading to glycogen depletion this mechanism will lead to fatigue, a
shortage of 2-oxoglutarate and glutamate and ammonia accumulation. BCOA =
Branched-chain 2-oxo acids; PEP = phosphoenolpyruvate.

nontoxic amino group carrier released by muscle. Alternatively, glutamate
may donate the amino group to pyruvate (derived from glucose or glycogen
oxidation) to form alanine and regenerate 2-oxoglutarate in the alanine
aminotransferase reaction (fig. 1). It is quite likely that the draining effect of
the BCAA aminotransferase reaction on the citric acid cycle is normally
counteracted by the anaplerotic carbon flux from glycogen, glucose and

pyruvate into the citric acid cycle (fig. 1). Glycogen and glucose, thus, also may provide carbon skeletons for glutamine synthesis. Pathways for the production of citric acid cycle intermediates from glycogen and glucose in muscle may involve carboxylation of pyruvate [15] and reversal of the phosphoenolpyruvate carboxykinase reaction [16]. Conversion of the carbon skeletons of valine and isoleucine to citric acid cycle intermediates also may help to counteract the draining effect of the BCAA aminotransferase reaction and to supply carbon skeletons for glutamine synthesis [10]. During exercise muscle, therefore, normally releases large amounts of glutamine and alanine in excess of its occurrence in muscle proteins and relatively small amounts of ammonia [13].

In the patient with McArdle's disease glycogen cannot be broken down to pyruvate during exercise. This may lead to a twofold problem with regard to disposal mechanisms for amino groups and ammonia. Firstly, less pyruvate is available as an amino group acceptor in the alanine aminotransferase reaction and secondly the anaplerotic carbon flux from glycogen to the citric acid cycle is disturbed (fig. 1). This may lead to a net drain on the citric acid cycle by the BCAA aminotransferase reaction, a reduced flux in the citric acid cycle and a limitation of aerobic oxidation of the available fuels. This may also lead to reduced availability of glutamate in the glutamine synthase reaction and therefore to accumulation of ammonia. Accumulation of ammonia in McArdle's disease can thus be seen as a biochemical marker of fatigue caused by a relative shortage of citric acid cycle intermediates.

Evidence in support of the presence of this fatigue mechanism in McArdle's disease comes from oral supplementation studies with BCAA and their 2-oxo acid analogues (BCOA). Supplementation with BCAA deteriorated exercise performance (increased heart rate and increased perceived exertion on the Borg scale) during incremental cycle ergometry and increased the plasma ammonia concentration during exercise, while supplementation with BCOA improved exercise performance and reduced the increase of the plasma ammonia concentration [10]. Glucose (75 g) given orally before incremental exercise also delayed exhaustion and reduced the increase in plasma ammonia concentration, heart rate and perceived exertion at a given power output [10]. The effects of glucose supplementation may be caused by a better provision of fuel and by anaplerotic provision of citric acid cycle intermediates and glutamate. In support of the suggested fatigue mechanism also is the observed positive correlation between heart rate and plasma ammonia concentration during incremental exercise (fig. 2). This correlation is not an epiphenomenon since it was power output independent.

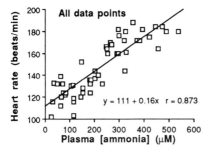 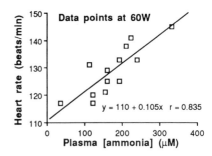

Fig. 2. The relationship between plasma ammonia concentration and heart rate during 13 incremental cycling exercise tests in a patient with McArdle's disease. The initial power output was 40 W increasing at 6-min intervals by 20 W until exhaustion. Heart rate was recorded and forearm venous plasma ammonia was measured as described [17] every 6 min and at exhaustion. In some tests the patient received supplements of BCAA, BCOA, glucose and BCOA+glucose. In the left panel all combined data points are given. In the right panel only data points obtained after 6 min at 60 W.

Another important indication for the existence of this fatigue mechanism and for the role of glycogen as a precursor for the synthesis of the carbon skeletons of glutamate and glutamine is the observation that patients with electron transport chain defects have an even greater intolerance to exercise than patients with McArdle's disease but do not produce excessive amounts of ammonia in ischemic forearm exercise [Lewis, personal commun.].

Role of Amino Acids and Ammonia in Fatigue Mechanism in Healthy Individuals

When the proposed fatigue mechanism applies to patients with a glycogen breakdown defect, then it is to be expected that it will also apply to healthy individuals in the glycogen-depleted state. Previously, we have shown that exercise in the glycogen-depleted state leads to a greater activation of the BC-complex in muscle than exercise following carbohydrate loading [18]. Blood ammonia concentration rose continuously during prolonged cycle ergometer exercise leading to glycogen depletion [11, 19]. The increase in plasma ammonia concentration is also greater in healthy individuals during exercise in the glycogen-depleted state (fig. 3) as compared with the undepleted state.

When BCAA metabolism is accelerated during exercise and thereby draining the citric acid cycle as proposed in figure 1, then one would expect

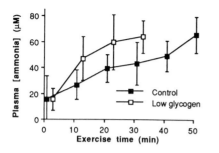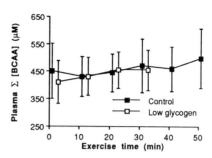

Fig. 3. Seven healthy individuals were studied during cycling exercise with and without a prior partial glycogen-depletion protocol. The muscle glycogen concentration was decreased by an exercise protocol at alternating intensities as described previously [20]. After warming up subjects cycled in 2-min blocks alternating at 90 and 50% W max. This was continued until the 2-min block at 90% W max could not be completed anymore. The high intensity was then reduced to 80% until it could not be completed anymore and finally to 70% W max. When the 70% W max block could not be completed the subjects were allowed to rest for 2 h. The exercise test itself consisted of 30 min of cycling exercise at 65% W max followed by exercise at 75% W max until exhaustion. Bloods were drawn from a forearm venous cannula, plasma spun and frozen for analysis of ammonia [17] and amino acids [24].

changes in the muscle concentration of 2-oxoglutarate, glutamate, glutamine, alanine and ammonia. Graham and Saltin [21] studied 6 male volunteers during brief intense cycle ergometer exercise to exhaustion. At exhaustion muscle 2-oxoglutarate fell to 52% of rest level, muscle glutamate fell to 30% of rest level and muscle ammonia doubled from about 300–600 μmol/kg wet muscle. Bergström et al. [22] studied 4 male volunteers during 20 min of bicycle exercise at 70% $\dot{V}_{O_2}$ max. Muscle biopsies were obtained at rest, after 10 min and after 20 min. Muscle glutamate decreased from 4.25 mmol · L^{-1} intracellular water at rest to values of 1.61 and 1.48 after 10 and 20 min of exercise. Muscle alanine increased from 2.36 mmol L^{-1} at rest to a value of 3.73 after 10 min of exercise and then returned to a value of 3.23 after 20 min of exercise. Muscle glutamine increased from 18.9 mmol · L^{-1} at rest to a value of 23.6 after 10 min of exercise and then returned to a value of 20.9 after 20 min of exercise. The observed increase in the glutamine concentration after 10 min of exercise (+4.7 mmol · L^{-1}) most likely originates from production of glutamine in the glutamine synthase reaction. This requires an amount of ammonia which is much greater than changes observed in the concentration of adenine nucleotides during this type of exercise. This would imply that ammonia production in healthy individuals not only originates from the breakdown of adenine nucleotides to IMP but also from the metabolism of amino acids as suggested before for the patient with McArdle's disease [10]. In many previous investigations, reviews and discussions on the origin of the muscular ammonia production during exercise and the functioning of the purine nucleotide cycle the amount of ammonia bound in the form of glutamine was not taken into consideration. The increase in the muscle glutamine concentration observed after 10 min of exercise [22] is also greater than the decrease in muscle glutamate. This implicates that alternative carbon precursors (e.g. glycogen, glucose, valine and isoleucine) are apparently used for the synthesis of glutamine. During exercise at 80% $\dot{V}_{O_2}$ max human muscle furthermore releases substantial amounts of glutamine and alanine to the circulation [13]. The increase in muscle glutamine concentration observed by Bergström et al. [22] hase been disputed more recently by Katz et al. [14], who studied 8 male volunteers also during brief intense exercise to exhaustion. Muscle glutamine remained at rest level in their study, muscle alanine doubled at exhaustion, muscle glutamate fell to $<25\%$ of rest level and muscle ammonia increased eight-fold. Settlement of the discrepancy with respect to muscle glutamine concentrations during brief intense exercise is an important issue in the discussions on the sources of ammonia production during exercise, the functioning of the purine nucleotide cycle in muscle and the presence of the alternative mechanism of fatigue suggested in this paper. Rennie et al. [23] studied 4 male volunteers during 3.75 h of treadmill exercise at 50% $\dot{V}_{O_2}$ max. Muscle glutamate fell from 4.75 to 2.80 mmol $\cdot$ L^{-1} intracellular water, while muscle glutamine decreased from 21.6 to 14.3 mmol $\cdot$ L^{-1} and muscle alanine remained unchanged at about 3.0 mmol $\cdot$ L^{-1}. Endurance exercise as used by Rennie et al. [23] will lead to glycogen depletion and a possible shortage of carbon precursors for glutamine and alanine synthesis. Most of these experimental data appear to be compatible with the existence of the proposed fatigue mechanism and the role of glycogen, glucose and pyruvate as carbon precursors for glutamate and glutamine synthesis.

To find further evidence of the presence of the alternative fatigue mechanism we also gave supplements of leucine and of the three BCAA to healthy individuals prior to exercise testing in the glycogen-depleted state (table 1). In contrast to patients with McArdle's disease time to exhaustion was not influenced by these supplements in comparison with the control test. The BCAA, however, increased heart rate at exhaustion (table 1). BCAA and leucine both increased plasma ammonia and plasma glutamine concentrations at exhaustion (table 1).

Table 1. Effect of leucine and BCAA supplements on exercise performance and plasma concentrations of ammonia and amino acids in glycogen-depleted healthy individuals

	Control test	+Leucine	+BCAA
Time to exhaustion, min	31.2 ± 6.6	32.7 ± 6.1	34.9 ± 5.8
Heart rate, beats · min^{-1}	174 ± 12	174 ± 10	181 ± 12*
Plasma (ammonia), μM			
Start	15 ± 10	33 ± 8*	23 ± 14
Exhaustion	68 ± 33	201 ± 87**	139 ± 55**
Plasma (glutamine), μM			
Start	533 ± 120	681 ± 178*	653 ± 166*
Exhaustion	588 ± 90	760 ± 186*	754 ± 137*
Plasma Σ(BCAA)			
Start	423 ± 118	1,953 ± 740**	2,988 ± 1,015**
Exhaustion	366 ± 130	1,435 ± 568**	2,446 ± 775**

Seven healthy individuals were studied during cycling exercise following a partial glycogen-depletion protocol. The muscle glycogen concentration was decreased by an exercise protocol at alternating intensities as described in the legend of figure 3. Leucine (0.35 g/kg body weight) and a mixture of the three BCAA (0.12 g/kg body weight of leucine, isoleucine and valine) were given orally suspended in skimmed yoghurt (2 g/kg body weight) 30 min after the glycogen-depletion protocol. In the control experiment the subjects only received yoghurt. The exercise test was started 2 h after the glycogen depletion test and consisted of 30 min (or time to exhaustion) of cycling exercise at 65% W max followed by exercise at 75% W max until exhaustion. Bloods were drawn at exhaustion from a forearm venous cannula, plasma spun and analyzed for ammonia [17] and amino acids [24]. Values are given as means ± SD. Significantly different from corresponding value in control test at *p < 0.05 and **p < 0.01 (Wilcoxon signed rank test).

All these data taken together appear to indicate that the alternative mechanism of fatigue is also present in healthy individuals but that it cannot be induced as easily by oral BCAA supplements as in patients with McArdle's disease. The main reason for this may be that the situation in McArdle's disease is much more extreme. In McArdle's disease the carbon flux from muscle glycogen to the citric acid cycle will be zero (fig. 4). In glycogen-depleted healthy individuals the carbon flux from glycogen will be gradually reduced but will never be zero since it is virtually impossible to reduce muscle glycogen below 60–70 μmol · g^{-1} dry muscle. Furthermore, during the first minutes of exercise hepatic gluconeogenesis and glycolysis will be slow in McArdle's disease, so that the uptake of glucose from the circulation will be

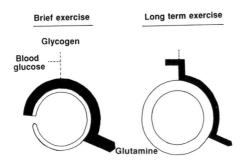

Brief exercise **Long term exercise**

Glycogen

Blood glucose

Glutamine

Fig. 4. Schematic presentation of the carbon fluxes feeding and draining the citric acid cycle and the consequences for the carbon flux circulating in the citric acid cycle during brief and long-term exercise in McArdle's disease. The carbon fluxes from glycogen and glucose enter the citric acid cycle at oxaloacetate, wheras draining occurs by the BCAA aminotransferase reaction (see fig. 1). The width of the arrow is proportional to the size of the flux.

small and also the carbon flux from blood-borne glucose into the citric acid cycle will be small (fig. 4). In the healthy individual endurance exercise will only gradually deplete muscle glycogen, and hepatic glycolysis and gluconeogenesis will be switched on simultaneously as a compensatory mechanism (fig. 5). Activation of the BC-complex is extremely rapid in McArdle's disease so that there is an immediate net drain of the BCAA aminotransferase reaction on the citric acid cycle during exercise (fig. 4). In healthy individuals activation of the BC-complex in endurance exercise is much more gradual, becoming more substantial when the carbohydrate stores are shrinking. Activation of the BC-complex may be much more rapid during, for example, sprint exercise. However, most of the energy provision in that case will come from anaerobic substrate oxidation and draining of the citric acid cycle flux will not be relevant for exercise performance. The observed changes in plasma and muscle metabolites nevertheless seem to indicate that similar metabolic reactions are activated as in endurance exercise and give rise to ammonia production. In this respect it also is important to realize that depletion of muscle 2-oxoglutarate and glutamate which is observed during brief intense exercise may impede the malate-aspartate shuttle and thus prevent the oxidation of cytosolic reduction equivalents [2]. Activation of branched-chain amino acid metabolism may thus also gradually reduce the percentage of energy which can be derived from aerobic oxidation during brief intense exercise (fig. 5) and thus also play a role in the development of fatigue in brief intense exercise.

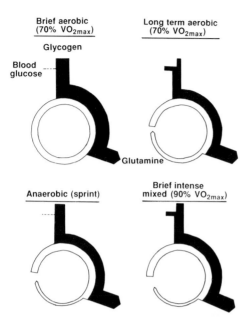

Fig. 5. Schematic presentation of the carbon fluxes feeding and draining the citric acid cycle and the consequences for the carbon flux circulating in the citric acid cycle during brief and long-term aerobic exercise at 70% $\dot{V}_{O_2}$ max and during brief intense anaerobic and mixed (90% $\dot{V}_{O_2}$ max) exercise in healthy individuals. For other details, see legend to figure 4.

The proposed fatigue mechanism also appears to explain why healthy individuals will experience a gradually increasing amount of exertion during prolonged exercise at constant power output leading to glycogen depletion, while patients with a glycogen breakdown defect experience a first phase with a peak in exertion followed by a second phase in which the pain and exertion ease (second wind [25]). During the first minutes of exercise there will be a net drain on the citric acid cycle in McArdle's disease, since muscle glycogen cannot be broken down and glycogenolysis and gluconeogenesis in the liver are not yet activated (fig. 4). In agreement with the proposed mechanism, the peaks in heart rate and perceived exertion experienced by patients with McArdle's disease during the first minutes of exercise at constant power output are followed within 2–3 min by a peak in the plasma ammonia concentration (fig. 6) [10].

The mechanism of fatigue proposed above should not be regarded as a separate entity but most likely is part of the general concept of fatigue. It is

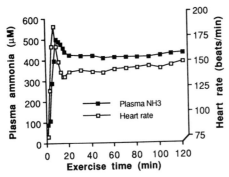

Fig. 6. A patient with McArdle's disease exercised for 2 h at 60 W on a cycle ergometer. Heart rate was monitored continuously. Bloods were drawn from a forearm venous cannula, plasma spun and frozen for analysis of ammonia [17].

quite likely that metabolic control of the citric acid cycle flux is part of a concerted action leading to respiratory control in general and to a balance between fuel availability and fuel oxidation and also between ATP production and ATP consumption. It is quite likely therefore that the activation of the BC-complex during exercise is under control of an important cellular modulator playing a major role in respiratory control in general and in the development of fatigue. ^{31}P-NMR investigations in patients with McArdle's disease have shown substantial decreases of the creatine phosphate concentration and an increase of P_i during exercise [26, 27]. Since the intramuscular pH is changing only marginally during exercise in McArdle's disease [26, 27] one may deduce from the creatine kinase equilibrium (creatine phosphate $+ADP_f+H^+\rightleftharpoons$ creatine$+$ATP) that the concentration of free ADP (ADP_f, not bound to protein) should increase. ^{31}P-NMR studies in a patient with McArdle's disease [27] indeed indicated that the concentration of ADP_f rose rapidly during the early stages of exercise at constant power output when the patient was experiencing most muscle pain and exertion and fell again during the second wind when the pain and exertion eased. In healthy individuals the increase in muscle ADP_f was much smaller than in the patient [27]. ADP_f is also a cytosolic modulator of respiratory control and its concentration rises much slower during exercise in trained than in normal muscle [28]. ADP may inhibit the myosin ATPase responsible for the actin-myosin interaction [29], the Ca^{2+}-ATPase of the sarcoplasmic reticulum involved in muscle relaxation, and the Na^+, K^+-ATPase of the sarcolemma involved in muscle activation [30]. Lau et al. [31] have identified ADP as an inhibitor of the kinase of the BC-complex and ADP_f therefore seems a likely common modulator of respiratory control, activation of branched-chain amino acid metabolism, the maximal aerobic rate of fuel oxidation, failure of the contractile machinery and fatigue.

Fatigue may be considered as a safety mechanism. It prevents dramatic changes in metabolism that could result in irreversible damage to muscles and even to other organs such as the brain. The draining action of the BCAA aminotransferase reaction thereby can be seen as a mechanism which prevents a disproportionate glucose uptake and oxidation in muscle and the development of hypoglycemia once the liver and muscle glycogen stores are shrinking. If, in the glycogen-depleted state, the rate of glucose uptake by muscle is greater than the difference between the rate of gluconeogenesis and the glucose requirement of nonmuscle tissues then the blood glucose concentration will fall. This will slightly reduce the muscle uptake and oxidation of glucose and lead to a small increase in the ADP_f concentration. This will activate BCAA metabolism, reduce the citric acid cycle flux and therefore reduce fuel oxidation by muscle. In some way other cellular events (K^+, Ca^{2+}, P_i or ADP_f) in parallel will lead to failure of the contractile machinery. The runner therefore will reduce pace until a new balance has been reached. Once the liver and muscle glycogen stores have been emptied then it is quite likely that the rate of gluconeogenesis may set the limit for the rate of synthesis of new citric acid cycle intermediates in muscle and therefore for the maximal rate of aerobic oxidation. This rate may well be the 50% of maximal aerobic oxidation observed during ultradistance running in healthy individuals. In this respect it is also interesting to note that the maximal aerobic power output in patients with McArdle's disease is about 50% of what would be expected from a healthy individual of similar sex, built and age.

Role of Amino Acids and Ammonia in Mechanisms of Central Fatigue

Investigations of the cerebral metabolism, locations, movements and electrical activities of the amino acids, of their metabolites, the amines, and of oligopeptides are beginning to dominate the fields of neurochemistry and neurophysiology. The reason for this eruption of research activity in the last two or three decades is the realization that the cerebral amino acids and their immediate derivatives influence and control mental behavior. Exercise in extremis may lead to loss of motor control and coherent thought. It is therefore no surprise that investigations have started and hypotheses have been put forward on the potential role of amino acids in mechanisms of central fatigue.

Newsholme and colleagues [4, 6, 7] have hypothesized that the free (= not bound to albumin) tryptophan concentration and changes in the tryptophan/BCAA blood ratio are involved in the production of central fatigue during prolonged exercise. BCAA and tryptophan enter the brain upon the same amino acid carrier and therefore compete for entry into the brain. Newsholme [4] reported that the blood concentration of BCAA decreased in human subjects during the marathon. This will increase the ratio of tryptophan to branched-chain amino acids in the bloodstream and favor the entry of tryptophan into the brain. Tryptophan is converted in the brain to a neurotransmitter known as 5-hydroxytryptamine (5-HT or serotonin). Evidence has been presented that tiredness and sleep may be influenced by the brain level of 5HT and therefore the link to central fatigue has been made [4, 6, 7]. A further important point in this hypothesis is that fatty acids compete with tryptophan for binding to albumin. During prolonged exercise leading to glycogen depletion the plasma fatty acid level will rise and increase the concentration of free tryptophan. All the described changes in blood concentrations of amino acids have been shown to occur during prolonged exercise in man [4] and rat [6, 7]; in rats endurance exercise has also been shown to raise the concentration of 5-HT in two areas of the brain [7]. However, it has not been shown that increased concentrations of 5-HT indeed cause central fatigue.

According to this hypothesis the effect of giving BCAA supplements to healthy individuals would be an improvement of performance (note that we had expected a deterioration of performance on the basis of the muscle fatigue mechanism). Our data in table 1 show that BCAA supplements increase the plasma BCAA levels about sevenfold but do not change performance. The substantial increase in the plasma ammonia concentration (table 1) can be seen as a contraindication for use of BCAA supplements by athletes. Figure 3 furthermore shows that endurance exercise leading to exhaustion not always leads to a decrease in the plasma concentration of BCAA as observed by Newsholme [4] during the Stockholm marathon. Another implication of the hypothesis would be that supplements of tryptophan would lead to central fatigue and deteriorate performance. Again this does not seem to be the case. Segura and Ventura [32] reported that tryptophan supplements improved performance during running exercise at 80% $\dot{V}_{O_2}$ max and it was claimed that this was due to an increased pain tolerance as a result of *L*-tryptophan ingestion. We feel that great care should be taken with *L*-tryptophan. If it increases pain tolerance then it bypasses the safety mechanism of fatigue and its use might lead to irreversible damage during competitive sport events. Furthermore, our knowledge of the physiology of this 5-HT precursor during exercise is too limited to know whether its use is safe and which doses are acceptable. Severe complications have been observed due to the use of either too high doses or an impure preparation of tryptophan [33].

Banister and co-workers [8, 9] have suggested that exercise-induced hyperammonemia may play a significant role in the development of central fatigue and that during exercise in extremis serious hyperammonemia may occur similar to that leading to loss of motor control, coherent thought and consciousness in pathological states. However, this claim has not been underpinned by data actually showing that the plasma ammonia concentration may increase during exercise in healthy individuals to levels comparable with pathological states. To the best of our knowledge the highest plasma ammonia concentrations which ever have been described in endurance exercise in man are about 250 μM [34]. These values were reached for short transient periods only during and following a 5-hour cycling exercise protocol at alternating intensities. Most of the subjects developed muscle cramps when these values were reached and complained of severe exertion and pain in their muscles. However, no signs of motor incoordination, ataxia or stupor could be observed. Most times peripheral fatigue has led to a substantial reduction of the pace before athletes start to demonstrate signs of central nervous system dysfunction, ataxia and stupor and increases in plasma ammonia concentration are not maintained when power output has to be reduced to levels below 70% W max because of fatigue and exhaustion [34]. At constant power output a patient with McArdle's disease could easily exercise for 2 h with plasma ammonia concentrations of $> 400\ \mu M$, a heart rate of 140–150 beats · min^{-1} and a rated perceived exertion on the Borg-scale of 12–14 (fig. 6). No signs of central nervous system dysfunction could be observed in this patient. It therefore seems unlikely that exercise-induced hyperammonemia plays a role in the development of ataxia and stupor during exercise in extremis and no data appear to underpin a role of exercise-induced hyperammonemia in central mechanisms of fatigue.

An important point which should be considered with respect to the relevance of central fatigue mechanisms as compared to local muscle fatigue mechanisms in endurance exercise is the following. In healthy, well-motivated subjects sustained maximal voluntary contractions can lead to $> 50\%$ loss of force within 1 min. There is good evidence that this type of fatigue occurs in the muscle itself rather than within the central nervous system or at the neuromuscular junction since maximum voluntary force generation falls in parallel with the force from electrically stimulated contractions [35]. One can easily imagine that a similar type of local fatigue mechanism may reduce the pace of the athlete by 50% in endurance exercise. To the best of our knowledge central fatigue, caused by modulation of neurotransmitters or ammonia, has never been shown to lead to a loss of force in isolated muscles studied in vivo.

Acknowledgments

Studies in patients with McArdle's disease have been funded by grants of the Muscular Dystrophy group of Great Britain and Northern Ireland, the Mersey Regional Health Authority, and Sigma Tau Spa, Pomezia, Italy. Studies in healthy subjects have been funded by an Isostar Research Grant of Wander Ltd., Bern, Switzerland. Degussa AG, Hanau, FRG, is gratefully acknowledged for the supply of infusion quality amino acids. I wish to thank my colleagues in Liverpool and Maastricht for advice, help and encouragement during the course of these studies.

References

1 Edwards RHT: Human muscle function and fatigue; in Porter R, Whelan J (eds): Human Muscle Fatigue: Physiological Mechanisms. Ciba Found Symp London, Pitman, 1981, vol 82, pp 1–18.
2 Newsholme EA, Leech AR: Biochemistry for the Medical Sciences. Chichester, Wiley, 1983.
3 Newsholme EA: Application of metabolic logic to the questions of causes of fatigue in marathon races; in Macleod D, Maughan R, Nimmo M, Reilly T, Williams C (eds): Exercise: Benefits, Limits and Adaptations. London, Spon, 1987, pp 181–198.
4 Newsholme EA: Metabolic causes of fatigue in track events and the marathon; in Benzi G (ed): Advances in Myochemistry, part 2. Montrouge, Libbey Eurotext, 1989, pp 263–271.
5 Goldberg AL, Chang TW: Regulation and significance of amino acid metabolism in skeletal muscle. Fed Proc 1978;37:2301–2307.
6 Acworth I, Nicholass J, Morgan B, Newsholme E: Effect of sustained exercise on concentrations of plasma aromatic and branched-chain amino acids and brain amines. Biochem Biophys Res Commun 1986;137:149–153.
7 Blomstrand E, Perrett D, Parry-Billings M, Newsholme EA: Effect of sustained exercise on plasma amino acid concentrations and on 5-hydroxytryptamine metabolism in six different brain regions of the rat. Acta Physiol Scand 1989;136:473–481.
8 Mutch BJC, Banister EW: Ammonia metabolism in exercise and fatigue: A review. Med Sci Sports Exerc 1983;15:41–50.

9   Banister EW, Cameron BJC: Exercise-induced hyperammonemia: Peripheral and central effects. Int J Sports Med 1990;11:S129–S142.

10 Wagenmakers AJM, Coakley JH, Edwards RHT: Metabolism of branched-chain amino acids and ammonia during exercise: Clues from McArdle's disease. Int J Sports Med 1990;11:S101–S113.

11 Wagenmakers AJM, Brookes JH, Coakley JH, Reilly T, Edwards RHT: Exercise-induced activation of the branched-chain 2-oxo acid dehydrogenase in human muscle. Eur J Appl Physiol 1989;59:159–167.

12 Wahren J, Felig P, Havel RJ, Jorfeldt L, Pernow B, Saltin B: Amino acid metabolism in McArdle's syndrome. N Engl J Med 1973;288:774–777.

13 Eriksson LS, Broberg S, Björkman O, Wahren J: Ammonia metabolism during exercise in man. Clin Physiol 1985;5:325–336.

14 Katz A, Broberg S, Sahlin K, Wahren J: Muscle ammonia and amino acid metabolism during dynamic exercise in man. Clin Physiol 1986;6:365–379.

15 Davis EJ, Spydevold Ø, Bremer J: Pyruvate carboxylase and propionylCoA carboxylase as anaplerotic enzymes in skeletal muscle mitochondria. Eur J Biochem 1980; 110:255–262.

16 Krebs HA: The role of chemical equilibrium in organ function. Adv Enzyme Regul 1975;15:449–472.

17 Janssen MA, van Berlo CLH, van Leeuwen PAM, Soeters PB: The determination of ammonia in plasma and whole blood; in Soeters PB, Wilson JHP, Meijer AJ, Holm E (eds): Advances in Ammonia Metabolism and Hepatic Encephalopathy. Amsterdam, Excerpta Medica, 1988, pp 587–592.

18 Wagenmakers AJM, Beckers E, Brouns F, Kuipers H, van der Vusse G, Saris WHM: Effect of carbohydrate availability on exercise induced activation of the branched-chain 2-oxo acid dehydrogenase in human muscle. Med Sci Sports Exerc 1989;21: S106.

19 Graham TE, Pedersen PK, Saltin B: Muscle and blood ammonia and lactate responses to prolonged exercise with hyperoxia. J Appl Physiol 1987;63:1457–1462.

20 Kuipers H, Keizer HA, Brouns F, Saris WHM: Carbohydrate feeding and glycogen synthesis during exercise in man. Pflügers Arch 1987;410:652–656.

21 Graham TE, Saltin B: Estimation of the mitochondrial redox state in human skeletal muscle during exercise. J Appl Physiol 1989;66:561–566.

22 Bergström J, Fürst P, Hultman E: Free amino acids in muscle tissue and plasma during exercise in man. Clin Physiol 1985;5:155–160.

23 Rennie MJ, Edwards RHT, Krywawych S, Davies CTM, Halliday D, Waterlow JC, Millward DJ: Effect of exercise on protein turnover in man. Clin Sci 1981;61:627–639.

24 Van Eijk HMH, van der Heijden MAH, van Berlo CLH, Soeters PB: Fully automated liquid chromatographic determination of amino acids. Clin Chem 1988;34:2510–2513.

25 Pearson CM, Rimer DG, Mommaerts WFHM: A metabolic myopathy due to absence of muscle phosphorylase. Am J Med 1961;30:502–517.

26 Lewis SF, Haller RG, Cook JD, Nunnally RL: Muscle fatigue in McArdle's disease studied by ^{31}P-NMR: effect of glucose infusion. J Appl Physiol 1984;57:1749–1753.

27 Radda GK: The use of NMR spectroscopy for the understanding of disease. Science 1986;233:640–645.

28 Dudley GA, Tullson PC, Terjung RL: Influence of mitochondrial content on the sensitivity of repiratory control. J Biol Chem 1987;262:9109–9114.

29   Bendall JR: A study of the kinetics of the fibrillar adenosine triphosphatase of rabbit skeletal muscle. Biochem J 1961;81:520–535.

30   Trentham DR, Eccleston JF, Bagshaw CR: Kinetic analysis of ATPase mechanisms. Q Rev Biophys 1976;9:217–281.

31   Lau KS, Fatania HR, Randle PJ: Regulation of the branched-chain 2-oxoacid dehydrogenase kinase reaction. FEBS Lett 1982;144:57–62.

32   Segura R, Ventura JL: Effect of L-tryptophan supplementation on exercise performance. Int J Sports Med 1988;9:301–305.

33   Slutsker L, Hoesley SC, Miller L, Williams LP, Watson JC, Fleming DW: Eosinophilia-myalgia syndrome associated with exposure to tryptophan from a single manufacturer. JAMA 1990;264:213–217.

34   Brouns F, Beckers E, Wagenmakers AJM, Saris WHM: Ammonia accumulation during highly intensive long-lasting cycling: individual observations. Int J Sports Med 1990;11:S78–S84.

35   Bigland-Ritchie B, Jones DA, Woods JJ: Excitation frequency and muscle fatigue: electrical responses during human voluntary and stimulated contractions. Exp Neurol 1979;64:414–427.

Dr. A.J.M. Wagenmakers, Nutrition Research Centre,
Department of Human Biology, University of Limburg, PO Box 616,
NL–6200 MD Maastricht (The Netherlands)

Marconnet P, Komi PV, Saltin B, Sejersted OM (eds): Muscle Fatigue Mechanisms in
Exercise and Training. Med Sport Sci. Basel, Karger, 1992, vol 34, pp 87–114

Metabolism and Performance in Exhaustive Intense Exercise; Different Effects of Muscle Glycogen Availability, Previous Exercise and Muscle Acidity[1]

Bengt Saltin, Jens Bangsbo, Terry E. Graham, Lars Johansen

August Krogh Institute, University of Copenhagen, Denmark

Introduction

The influence of diet on the metabolic response to long-term exercise and exercise capacity was studied by Christensen and Hansen [1]. They demonstrated that a carbohydrate-enriched diet resulted in a higher RQ and better performance when compared to exercise at the same intensity after days with a fat and protein diet. Some 30 years later, Bergström and colleagues [2, 3] used a similar protocol with the addition of muscle glycogen determinations. Performance time was closely related to the initial glycogen level in the muscles engaged in the exercise, i.e. fatigue and exhaustion were postponed in relation to availability of muscle glycogen at the start of the prolonged exercise. Further, at exhaustion, muscles were completely glycogen depleted. Pernow and colleagues [4–6] also elaborated on the theme of substrate supply to exercising muscles and demonstrated that shortage of blood-borne substrates like FFA and glucose had an influence on the metabolism and exercise performance as well.

The question has been raised as to whether initial muscle glycogen concentration plays a role for short-term exercise capacity and if the rate of glycogenolysis is affected. In studies using rats, large muscle glycogen stores were shown to induce higher glycogenolytic and glycolytic rates. In fast

[1] The original data presented in this article were obtained in studies supported by grants from Team Denmark and Danish Natural Science Foundation (11-7776).

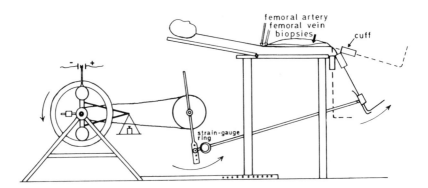

Fig. 1. An illustration of the experimental set up. Exercise ergometer. The ankle is attached via an aluminum bar to the crank of a Krogh bicycle ergometer. The exercise consists of kicking the lower part of one leg 'upwards'. The reposition of the leg is completely passive (i.e. gravity and momentum of the wheel).

twitch fibers glycogen utilization and lactate release were linearly related to the initial muscle glycogen concentration [7], findings which, according to Spriet et al. [8], were not due to a coupling between muscle glycogen and anaerobic glycogenolysis. In contrast, they found no relationship between muscle glycogen and anaerobic glycogenolysis. In man similar discrepancies are found in the literature [9–14].

The role of elevated acidity of the muscle for developing fatigue and causing exhaustion is another factor which has received much attention [15, 16]. Within the sports community it is taken as a fact that lactate accumulation and the concomitant increase in H^+ concentration is the cause of perceived fatigue and exhaustion in high intensity exercise. However, an exclusive role of H^+ concentration is not well established as neither variation in pre-exercise acidity nor muscle pH appear to be very critical [17–21].

Thus, several questions need to be addressed with regard to the metabolic response when performing intense all-out exercise. One is the role of initial muscle glycogen content and another the role of muscle lactate. There are different procedures to achieve variation in muscle glycogen and lactate content in man. For glycogen, dietary manipulation with or without previous exercise can be used. Another route is to lower the glycogen stores by exercise prior to the test exercise. We have used both approaches. To elevate muscle lactate, subjects exercised intensely and then repeated the work. The exercise model chosen was one-legged knee-extensor kicking (fig. 1), which allows for precise, quantitative metabolic evaluation [22], as well as relating these metabolic events directly to performed work and exhaustion [21]. Another advantage with the one-legged exercise method is that different interventions can be applied to the two limbs of the subjects, allowing for more direct comparison. The protocol made it possible to evaluate to what extent glycogenolysis and lactate production were functions of the initial muscle glycogen content and muscle lactate (muscle pH), and also to separate these effects from previous exercise. Further, the functional significance of an elevated muscle blood flow for release of lactate from muscle was also tested by performing light exercise during recovery.

Our results give clear answers to the questions raised. Above normal muscle glycogen content had little effect on the metabolic response and short-term exercise performance, whereas previous exercise had a definite effect, resulting in a lowering of glycolytic rate, lactate production and performance. This effect was exaggerated by elevated muscle lactate level, whereas 'active' recovery had minor effects on lactate release and performance. Particular details related to the problems discussed in this chapter can be found elsewhere [23–25].

Results

I. Above Normal Muscle Glycogen Content

To achieve a difference in the glycogen content of the knee-extensor muscle of the subject's two limbs, exercise with one leg was performed to exhaustion 3 days prior to the experimental day (high glycogen leg; HG). Thereafter, the subject had a carbohydrate-enriched diet. The evening prior to the experiment, kicking exercise was performed to fatigue with the other leg (control leg; C), whereafter no food intake was allowed, but water, ad libitum, was encouraged. On the day of the experiment, one leg performed high exercise intensity to exhaustion (which occurred after approximately 3 min) (fig. 2). After approximately 2 h of rest, the other leg was exercised at exactly the same exercise intensity. The dietary manipulation combined with the exercise preceding the experiment resulted in mean muscle glycogen contents of 87.0 and 176.8 mmol kg^{-1} w.w. ($p < 0.05$) in the C and HG leg, respectively (fig. 3). The exhaustive exercise caused a muscle glycogen depletion of 26.3 (C) and 25.6 (HG) mmol kg^{-1} w.w. ($p > 0.05$). Performance measured as time to exhaustion, averaged 2.82 (C) and 2.92 (HG) min ($p > 0.05$). Thus, over the range of initial muscle glycogen contents (from 45

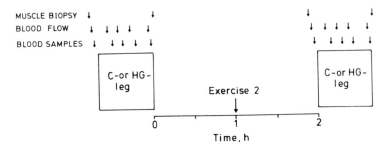

Fig. 2. Protocol for experiment I. Half of the subjects started with the 'control' (C) leg and half of them with the 'high muscle glycogen' (HG) leg. After 1 h of rest the subject repeated the exercise (exercise 2) with the same leg [modified from ref. 22].

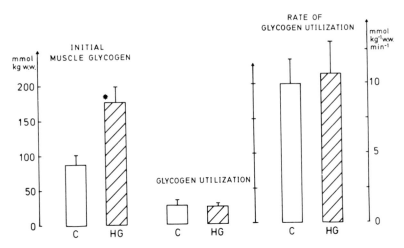

Fig. 3. Muscle glycogen concentrations before (left), glycogen utilization during (middle), and rate of glycogen utilization during (right) the exhaustive exercise bouts performed by the C leg and the HG leg, respectively [modified from ref. 22].

to just above 250 mmol kg^{-1} w.w.) the rate of glycogen utilization was unaffected (fig. 4) with the means being 10.0 (C) and 10.7 (HG) mmol kg^{-1} w.w. min^{-1} ($p > 0.05$) (fig. 3).

Production of lactate was also unaffected by the variation in initial muscle glycogen content, and insignificant ($p > 0.05$) variations were ob-

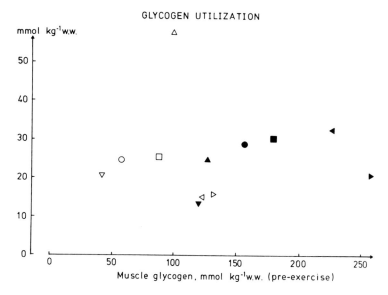

Fig. 4. Individual values for muscle glycogen concentration prior to the exhaustive exercise and glycogen utilization during the exhaustive exercise for the C leg (open symbols) and the HG leg (filled symbols) [modified from ref. 22].

served for the amount of lactate which accumulated in the muscle during the exercise, as well as the amount which was released (fig. 5). The former being 18.8 (C) and 16.1 (HG) mmol kg^{-1} w.w., and the latter 23.4 (C) and 26.5 mmol (HG) (p>0.05).

The relative role of aerobic and anaerobic energy yields during the exercise was the same for the two legs (fig. 6). Limb oxygen uptake accelerated at the same rate, and the same peak oxygen uptake was reached in each leg. Total leg oxygen uptake amounted to 1.51 (C) and 1.47 liters (HG), and the estimated oxygen deficit averaged 0.87 (C) and 0.91 (HG) liters 'O$_2$ equivalent (Eq.)' (p>>0.05). The latter had an initial rate of approximately 0.57 liters 'O$_2$ Eq.' min^{-1} during the exercise and 0.16 liters 'O$_2$ Eq.' min^{-1} during the final stage of the work.

Nucleotide and CP metabolism were also similar with an ATP depletion of 1.1 mmol kg^{-1} w.w. or around 20% reduction for both C and HG legs (table 1). IMP was elevated during the exercise to 0.68 (C) and 0.28 (HG) mmol kg^{-1} w.w., and CP was reduced by 12.2 (C) and 11.1 (HG) mmol kg^{-1} w.w., or by 64 and 58%, respectively.

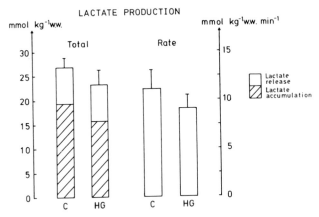

Fig. 5. Lactate production (left) and rate of lactate production (right) for the C leg and the HG leg (HG) during the exhaustive exercise [modified from ref. 22].

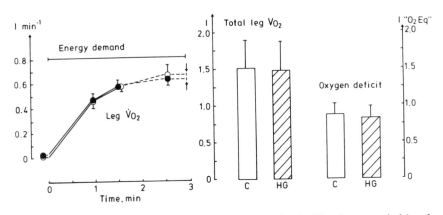

Fig. 6. Leg oxygen uptake and energy demand (left) for the C leg (open symbols) and the HG leg (filled symbols), and total leg oxygen uptake (middle) and oxygen deficit (right) for the C leg and the HG leg during the exhaustive exercise [modified from ref. 22].

A small, but significantly greater K^+ release was observed when the HG leg was exercising (fig. 7). The difference was already present in the first measurement (after approximately 45 s of work) and remained until exhaustion. The total release amounting to 4.4 (C) and 5.8 (HG) mmol ($p < 0.05$). It is of note that plasma flow (and blood flow, fig. 7) to the limb was identical in

Table 1. Muscle CP and nucleotide concentrations (mmol kg^{-1} w.w.) before (Pre) and after (Post) the exhaustive exercise bouts

| | Full recovery | | | | | | | | Short recovery | | | | | |
| | C | | HG | | EX1 | | EX2 | | EX1 | | EX2-HLa | | EX2-HLa-act. | |
	Pre	Post	Pre	Post	Pre	Post	Pre	Post	Pre	Post	Pre	Post	Pre	Post
CP	15.12 ±1.69	4.69 ±1.02	16.72 ±1.33	5.32 ±1.98	16.02 ±0.89	5.01 ±0.76	15.44 ±1.13	4.46 ±1.81	19.75 ±0.43	3.63 ±0.36	15.46 ±1.81	3.32 ±2.76	16.40 ±2.26	3.11 ±1.29
ATP	4.74 ±0.41	3.71 ±0.37	5.11 ±0.71	4.03 ±0.63	4.92 ±0.33	3.87 ±0.27	4.77 ±0.18	3.68 ±0.25	6.15 ±0.42	5.06 ±0.63	4.90** ±0.65	3.99** ±0.72	4.96* ±0.53	3.57* ±0.60
ADP	0.51 ±0.08	0.67 ±0.14	0.53 ±0.06	0.58 ±0.15	0.52 ±0.05	0.63 ±0.11	0.51 ±0.05	0.69 ±0.14	0.69 ±0.07	0.70 ±0.04	0.59 ±0.09	0.61 ±0.11	0.61 ±0.07	0.66 ±0.10
AMP	0.19 ±0.01	0.13 ±0.03	0.13 ±0.02	0.14 ±0.02	0.16 ±0.01	0.14 ±0.02	0.16 ±0.01	0.14 ±0.02	0.02 ±0.00	0.05 ±0.02	0.10** ±0.02	0.12** ±0.03	0.04 ±0.01	0.15* ±0.03
IMP	<0.01	0.68 ±0.09	<0.01	0.28 ±0.12	<0.01	0.53 ±0.13	<0.01	0.46 ±0.15	<0.01	0.42 ±0.14	0.38** ±0.16	0.51 ±0.21	0.13+ ±0.09	0.83 ±0.35

Means ±SE are given.
Significant difference (p<0.05) between: *EX1 and EX2-HLa-act., **EX1 and EX2-HLa, and +EX2-HLa and EX2-HLa-act.

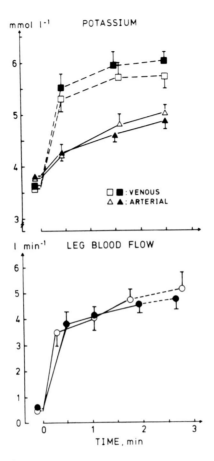

Fig. 7. Femoral arterial (triangles) and venous (squares) blood potassium concentrations (upper) and leg blood flow (circles) for the C leg (open symbols) and the HG leg (filled symbols) during the exhaustive exercise [modified from ref. 22].

the two exercise situations along with femoral arterial (4.98 and 4.85 mmol l^{-1}, respectively) and venous (5.70 and 6.00 mmol l^-) K^+ concentrations at exhaustion.

II. Previous Exercise

a. Single Bout and 'Full' Recovery. The experimental design described above included an evaluation of the effect of previous exercise, as each leg was exercised to exhaustion a second time (EX2). This was accomplished by

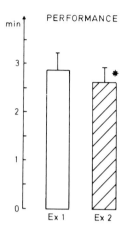

Fig. 8. Performance (time to exhaustion) in the EX1 and EX2 exercise bouts when full recovery was allowed (Exp. IIa) [modified from ref. 22]. *Significant difference (p < 0.05) between EX1 and EX2.

repeating the test exercise after one hour of recovery (fig. 2); meaning that the first limb exercised twice before the other leg became involved. As described above, the metabolic response to the exercise, except for the K^+ release (EX1) were quite similar regardless of the initial muscle glycogen content of the limbs. Thus, in evaluating the effect of previous exhaustive exercise, the data are pooled, i.e. the results of both legs are combined. The responses of the C and HG limbs were very similar when performing a second time (EX2), thus supporting the pooling of the data.

It is of note that resting muscle and blood lactate levels were allowed to return to pre-EX1 levels prior to the second exercise bout, and the muscle and arterial blood lactate concentrations were as low as 1.3 mmol kg^{-1} w.w. and 0.54 mmol l^{-1}, respectively.

Initial muscle glycogen content had been reduced by 25.9 mmol kg^{-1} w.w. in EX1, resulting in a range of muscle glycogen content from 32.8 to 227.7 with a mean of 106.0 mmol kg^{-1} w.w. at the start of EX2. The reduction of muscle glycogen in EX2 was 22.5 mmol kg^{-1} w.w., or 13% (p<0.05) less than during EX1. Time to exhaustion in EX2 was reduced with approximately 10% (p<0.05; fig. 8). Thus, the mean glycogen utilization rate was the same for EX1 and EX2 (10.4 and 10.2 mmol kg^{-1} w.w. min^{-1}, respectively).

Lactate production was markedly reduced (39%) to 15.3 mmol kg^{-1} w.w. in EX2 (p<0.05, fig. 9). This was true for lactate accumulation in muscle, as well as the lactate that was released during the exercise (fig. 9). The reduction

LACTATE PRODUCTION

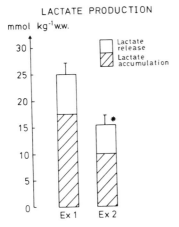

Fig. 9. Lactate production during the EX1 and EX2 exercise bouts in the same experiment as in figure 8 [modified from ref. 22].

being 43 and 28% ($p < 0.05$) for the two variables, respectively. The production rate for lactate was depressed, and amounted to 6.7 (EX2) as compared to 10.1 mmol kg^{-1} w.w. min^{-1} during EX1 ($p < 0.01$).

Rate of rise in aerobic energy yield as judged from limb oxygen uptake, was unaltered in EX2, but with a shorter exercise time, total oxygen consumed was reduced (14%, $p < 0.05$) along with total anaerobic energy production (7%, $p > 0.05$). The oxygen deficit was lowered from 0.88 (EX1) to 0.82 (EX2) liters · min^{-1} 'O$_2$Eq' ($p < 0.05$), matching the reduced lactate production observed when the leg exercised a second time. ATP and CP stores were utilized to the same extent during the two exercise bouts (table 1).

Potassium ion release during EX2 was similar to that occurring during EX1, with no difference between EX1 and EX2 in either arterial (4.91 and 4.95 mmol l^{-1}, respectively) or femoral venous plasma K$^+$ concentration (5.85 and 5.71 mmol l^{-1}, respectively) at exhaustion.

b. Repetitive Exercise Bouts and Short Recovery. In the above experiment, a recovery period was chosen so that the studied variables returned to pre-exercise levels prior to the second exercise test. In the second set of experiments, test exercises were the same as in the first study (EX1 and EX2; see fig. 2). The recovery period, however, was shortened to 16 min, and during this period seven bouts of high intensity exercise were performed

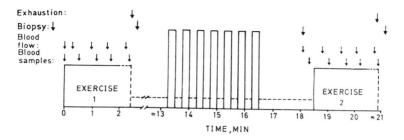

Fig. 10. Protocol for the experiment where muscle lactate was elevated prior to EX2 (Exp. IIb). The dotted horizontal line indicates the low intensity exercise during the recovery periods in the experiment with 'active' recovery (Exp. III) [modified from ref. 23].

(fig. 10). Thus, muscle lactate was 13.1 mmol kg^{-1} w.w. prior to the second test-exercise (EX2-HLa) and arterial blood lactate was 4.77 mmol l^{-1}. Prior to EX2 muscle pH was 6.85, which was lower (p<0.05) than before EX1 (7.04). At exhaustion from EX2-HLa muscle pH (6.77) tended to be higher (p<0.1) than at the end of EX1 (6.73) (fig. 11). Muscle glycogen had been depleted by 39.4 mmol kg^{-1} w.w. during the previous exercise (EX1 and the repetitive exercises) and was 94.1 mmol kg^{-1} w.w. prior EX2-HLa, with no leg of the subjects having a storage of less than 65.7 mmol kg^{-1} w.w.

The metabolic response during EX1 and EX2-HLa was markedly different, with a smaller reduction in muscle glycogen and less lactate produced in EX2-HLa (fig. 12). These responses were more extreme than those described above for EX2 after 1 h of recovery (IIa) where muscle lactate and acidity were not elevated prior to EX2. Glycogen depletion was reduced by 6.6 mmol kg^{-1} w.w., or 26% of EX1, this being about double the reduction observed with previous exercise and complete recovery. This pattern was similar for lactate production which was reduced by 18.7 mmol kg^{-1} w.w., or 55%, which was also twice as much as in EX2 of the previous study (IIa). Performance was impaired, the reduction being 45 s, or 20%, twice as much as in experiment IIa.

Limb oxygen uptake increased at the same rate during EX1 and EX2-HLa, and the same peak value was reached. However, due to the shorter exercise time the total leg uptake of 1.37 liters during EX2-HLa was 9% lower (p>0.05) than during EX1. The estimated oxygen deficit for EX2-HLa was 22% (0.24 liters 'O$_2$Eq') lower (p<0.05) than for EX1, which represents

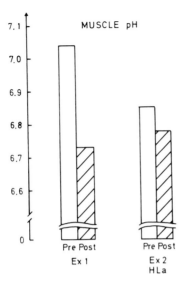

*Fig. 11.*Muscle pH before (Pre) and after (Post) the first (EX1) and the second (EX2-HLa) exhaustive exercises [modified from ref. 23].

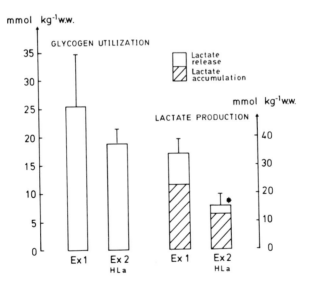

Fig. 12. Muscle glycogen utilization (left) and lactate production (right) during the first (EX1) and the second (EX2-HLa) exhaustive exercise [modified from ref. 23].

17 mmol ATP kg^{-1} w.w. or slightly less than the observed reduction in lactate production of 28 mmol ATP kg^{-1} w.w. Actual depletion of ATP and CP amounted to about 17 mmol kg^{-1} w.w. and was similar in EX1 and EX2-HLa, representing 23 and 28% of the total anaerobic energy production, respectively (table 1).

Changes in K$^+$ release during EX1 and EX2-HLa were similar, but the total release during EX2-HLa was 24% lower (p>0.05) than during EX1 (5.1 mmol), a result of the difference in exercise time. Arterial-venous K$^+$ difference at exhaustion was the same for EX1 (0.53 mmol l^{-1}) and EX2-HLa (0.66 mmol l^{-1}), as were the femoral venous concentrations similar (6.15 and 5.93 mmol l^{-1}, respectively).

III. Active Recovery

Low intensity exercise has been shown to accelerate lactate disappearance [26–30]. The suggested mechanisms include enhanced lactate utilization in the active muscle, increased lactate release from the muscle and increased blood lactate clearance.

The extent of such effects was evaluated by having the subjects repeat the protocol depicted in figure 10. However, instead of no exercise between work bouts, recovery was active, with the leg kicking at a modest intensity (13 W). Exercise at this intensity started immediately after EX1 and continued until the start of the second exhaustive exercise bout (EX2-HLa-act.), with the high intensity repetitive work bouts intervening. With this protocol, limb blood flow was elevated by 1.1 liters min^{-1} during recovery, compared to no exercise in the recovery (fig. 13). The release of lactate was 4.2 mmol min^{-1}, which was approximately 12% more (p<0.05) than observed in the experiments with 'passive' recovery. Muscle lactate prior to EX2-HLa-act. was 9.9 mmol kg^{-1} w.w., and arterial blood lactate 4.7 mmol l^{-1}. Muscle pH prior to EX2-HLa-act. was 6.84 as compared to 6.85 before EX2-HLa, and the reduction during the exercise amounted to 0.08 units, a change similar to experiment IIb. Further, muscle glycogen depletion prior to EX2-HLa-act. was 36.3 mmol kg^{-1} w.w.; however, 97.2 mmol kg^{-1} w.w. remained in the muscle, with no muscle having less glycogen stored than 78.0 mmol kg^{-1} w.w.

Muscle glycogen utilization by EX2-HLa-act. was depressed compared to EX1, with the reduction amounting only to 3.4 mmol kg^{-1} w.w. or 13% (p>0.05; fig. 14). Lactate production was smaller in EX2-HLa-act. compared to EX1, the difference in the order of 16.5 mmol kg^{-1} w.w. or 52% (p<0.05) being due to both less lactate accumulation and less released from the muscle during EX2-HLa-act. (fig. 14). These alterations were in the same range as

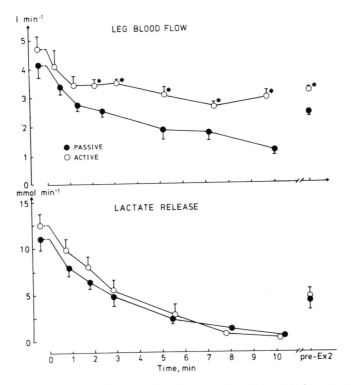

Fig. 13. Leg blood flow (upper) and lactate release (lower) during recovery with either low-intensity exercise (open symbols) or no (filled symbols) exercise between the two exhaustive exercises [modified from ref. 24]. *Significant difference (p < 0.05) between passive and active recovery.

described in EX2-HLa following passive recovery (IIb). Time to exhaustion tended to be lower than EX1 (3.46 vs. 3.00 min, 13%, p>0.05). The extent of this decrease was in between the results obtained in experiments IIa and IIb.

An effect of light exercise during recovery was that leg blood flow and oxygen uptake were at a higher level at the start of EX2-HLa-act. (fig. 15). This in turn affected the leg oxygen uptake and the rate of contribution of the anaerobic energy yield. Leg oxygen uptake during EX2-HLa-act. was slightly higher early in the exercise (p<0.05) than during EX1, but due to the shorter exercise time, total oxygen uptake was similar (1.59 and 1.67 liters, respectively). The initial rate of utilization of the oxygen deficit in EX2-HLa-act. was two-thirds that of EX1. Further, as leg oxygen uptake was high, the rate

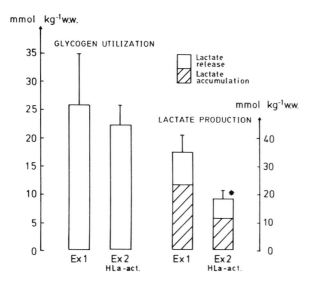

Fig. 14. Muscle glycogen utilization (left) and lactate production (right) during the first (EX1) and the second (EX2-HLa-act.) exhaustive exercises [modified from refs 23, 24].

of utilization of the anaerobic capacity continued to be low. This resulted in less than full usage of the capacity to accumulate an oxygen deficit, and was 0.26 liters 'O$_2$Eq' or 28% lower (p<0.05) than in EX1. The observed smaller oxygen deficit was related to a lower lactate production.

Also, in this part of the experiment, intervention caused ATP and CP depletion during the exercise, which were similar to those observed in the above-reported interventions (I, IIa and b), and amounted to 16.7 and 18.5 mmol kg^{-1} w.w. for EX1 and EX2-HLa-act., respectively (table 1).

The rate of K$^+$ release increased more rapidly during EX1, but was the same at exhaustion, with the total release of K$^+$ tending (p<0.1) to be higher in EX1 compared with EX2-HLa-act. (5.2 vs. 3.0 mmol). Both arterial-venous difference for K$^+$ (0.47 and 0.61 mmol l^{-1}, respectively) and femoral venous K$^+$ concentration (6.05 and 5.93 mmol l^{-1}, respectively) at exhaustion were the same.

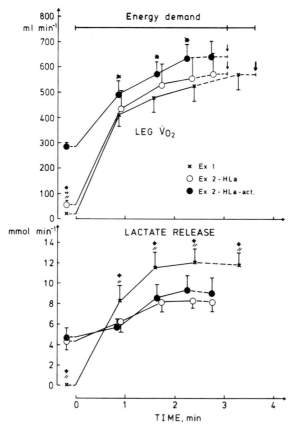

Fig. 15. Leg oxygen uptake and energy demand (upper), and lactate release during EX1 and EX2 exercises performed after intermittent exercise with either rest (passive, ○) or low intensity exercise (active, ●) in the recovery period [modified from refs 23, 24]. (*, //, ┿Significant difference (p < 0.05) between EX1 and EX2-HLa-act., EX1 and EX2-HLa and EX2-HLa and EX2-HLa-act., respectively.)

Discussion

A consensus has been reached that there is a close coupling between the magnitude of the carbohydrate store in the body and endurance exercise capacity. Thus, elongation of time to exhaustion at a given submaximal exercise intensity (<90% $\dot{V}O_2$ max) occurs even though the elevated CHO storage of the body often causes an elevation of 'aerobic glycolysis' and lactate production [1–3, 31, 32]. Further lipid oxidation is reduced.

In short-lasting heavy exercise, where the rate of 'anaerobic glycolysis' peaks, and its relative role to the energy yield is paramount, an even closer relationship between availability of glycogen in active muscle and its utilization has been proposed [7, 12]. Further, a scheme has been suggested where increased acidity of the muscle inhibits, at key regulatory sites, the rate of glycolysis and thus ATP resynthesis which induces fatigue and exhaustion [16]. These concepts have, however, far from full support by experimental data. The results presented in this chapter further amplify that the validity of these concepts is open for reappraisal.

The two main questions asked by us were: whether initial muscle glycogen content and the glycolytic rate in intense short-lasting exercise were coupled, and to what extent elevated muscle acidity affected metabolic responses and performance. The results presented in figures 3–5 give a straightforward answer to the first question. Glycolysis, neither as peak rate, nor as total contribution of energy during the exercise, are functions of above normal CHO storage in the muscles. Further, this relates both to its 'aerobic' and 'anaerobic' usage. The bulk of data in the literature on man are in line with our findings. Conflicting results are, however, also available. Asmussen et al. [33], and later Klausen and Sjögaard [12], studied the effect of dietary manipulation on lactate levels after short-lasting intense exercise. In the latter study, muscle analyses were also performed. The diets produced differences in carbohydrate storage of the muscles (from approximately 60 to 140 mmol · kg^{-1} w.w.). Muscle and blood lactate levels after repeated 2-min exhaustive exercise bouts were reduced in relation to the lowering of the muscle glycogen content. Of very special note is their finding of muscle glycogen depletion being identical (12.5 mmol · kg^{-1} w.w.) for each 2-min exercise bout (this is clearly pointed out by the authors in their discussion). This implies very similar rates of glycolysis. A possible explanation for the lower lactate concentration could therefore not be a reduced production, but rather an elevated utilization of the produced lactate in various tissues of the body when the CHO storage was scarce. Muscle glycogen levels were surprisingly high after the fat and protein diet in this particular study. It is likely, however, that glycogen in the various fibers was unevenly distributed, which was further pronounced after the first exercise bout, and resulted in elevating lactate turnover. In addition, it should be considered that several days with an extreme diet may enhance metabolic pathways for the use of lactate in oxidative metabolism [34]. Results in line with this chain of events are available. Jacobs and colleagues [11, 14, 35] have, in series of experiments, been unable to demonstrate over a large range of muscle glycogen contents an effect of the rate of glycogen depletion during short very intense exercise. There is one study where elevated muscle glycogen content results in an enhanced glycogenolysis and improved sprint performance [36, 37]. Further, the concentration of the intermediate in the glycolytic pathway was also increased, which could, via a Michelis-Menten 'effect', explain the higher rate of glycolysis. In this study the larger muscle glycogen storage was moderate and brought about by physical training. The observed findings are then more likely training induced rather than linked with alterations in the CHO storage. Our conclusion, therefore, is that above normal muscle glycogen content and glycolytic rate in intense exercise are not coupled. Differences in muscle and blood lactate concentrations, which are sometimes found, can be attributed to the lactate being metabolized at different rates.

It is more difficult to explain the findings on rat, which in part differ from man. Richter and Galbo [7] found that when using an isolated rat hind-limb model, and electrically inducing contractions for 15 min, particularly in fast-twitch oxidation (FO) and fast-twitch oxidation and glycolytic (FOG) fibers, a coupling existed between content of glycogen in muscle and the rate by which it was utilized. Spriet et al. [8] raised the question whether this was due to the well-documented coupling of aerobic consumption to carbohydrate in more prolonged exercise. They found in experiments where circulation was occluded, a pronounced drop in tension development with intermittent stimulation of the muscle within 200 s. However, glycogen depletion and rate of lactate production were similar regardless of initial muscle glycogen content. Thus, they concluded that the findings by Richter and Galbo [7] were compatible with their original suggestion [8]. Hespel and Richter [personal commun.] re-studied this problem using their experimental model, and included muscle samples after 1 and 2 min of stimulation. Both early and late in the stimulation period of 15 min, glycolysis was observed to be a function of the magnitude of the muscle glycogen storage. They attributed the linking of high muscle glycogen and high rates of glycolysis to a high content of phosphorylase a. Its activity at rest was positively related to the muscle glycogen content.

Electrically induced contractions mimic voluntary elicited muscle activity; however, there are certain definite differences. One is the recruitment pattern, which with electrical stimulation favors involvement of fast-twitch fibers [38]. This is probably not a serious concern in short-lasting experiments where fatigue develops within 1–2 min, as the FT fiber pool in voluntary contractions also contributes the most in these circumstances. In exercise lasting 15 min, as in the case of Richter and colleagues, a 'reversed' recruitment order is obtained which means that comparisons with voluntary contractions are not proper. In the former situation, the FT pool of fibers are engaged early in the exercise for force production but drop out quickly. For the remaining time, force production relates to ST fiber tension development. Thus, the two major types of muscle fibers are studied separately, with a major impact of FT muscle fiber metabolism during the 1- and 2-min samples, whereas slow-twitch (ST) muscle response is studied at the completion of exercise. This, by itself, is not an explanation for the finding of a linkage between glycogen content and its degradation. However, as a result of the supramaximal stimulation that was applied, it is likely that a high concentration of P_i develops, especially in fast-twitch muscles. Chassiotis and colleagues [39–41], and later Ren and Hultman [42], have shown that P_i is a critical activator of phosphorylase b to a, and thus rate of glycogen utilization. A likely explanation for the difference in results, comparing the findings of Richter and colleagues [7] with those of Spriet et al. [8] as well as those on man [21] could be this difference in the activation of muscle. Supramaximal activation may cause markedly higher free P_i levels in the contracting muscle.

A high rate of glycolysis will lead to the production of lactate, a reduced pH and elevated [H+]. There is no doubt that the increased acidity of the muscle has metabolic effects, and contributes to the development of fatigue. The questions are: How much, and site of action? The answer to the first question is that this effect is not very dramatic. At least not when taking into account the magnitude of elevation of muscle lactate and reduction in pH. The contribution of anaerobic glycolysis to the performed exercise is reduced, but the rate by which it is utilized appears to be less affected, which relates especially to its aerobic utilization. Of some interest in this respect are the data from Karlsson et al. [43] published some 20 years ago. Their subjects were characterized in regard to performance, glycogen depletion and lactate accumulation during ordinary bicycle exercise leading to exhaustion in 3–6 min. In separate experiments, exhaustive arm exercise was performed elevating plasma lactate concentration to 8–10 mM and leg skeletal muscle lactate level to ~ 6 mmol $\cdot$ kg^{-1} w.w. When leg exercise was then performed it did not affect performance [44], glycogen depletion or lactate production, i.e. muscle lactate concentration at exhaustion was elevated with the amount found in the muscle at the start of the exercise (~ 6 mmol $\cdot$ kg^{-1}). Without doubt these data add to the bulk of findings indicating that lactate accumulation in muscle only cause minor effects. It is a possibility, however, that lactate and [H+] transport out of the muscle cell may not follow each other exactly. As demonstrated by Juel et al. [17] and others [45, 46], the transport of these

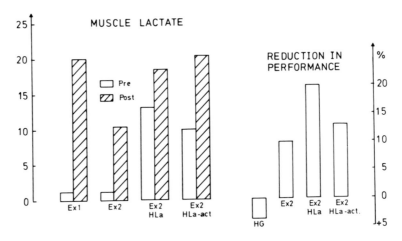

Fig. 16. Muscle lactate concentration (left) before (Pre) and after (Post) the first (EX1) and the second exhaustive exercise in the various interventions EX2; EX2-HLa; EX2-HLa-act. To the right is the alteration in performance for the HG leg, and for the EX2 bouts with full recovery and with elevation in muscle lactate in comparison to the control exercises [from refs 22–24].

have a different time course and in part are transported by separate mecha-nisms. Thus, in the experiment of arm exercise preceding the leg exercise [43, 44], the possibility exists that lactate is transported into the cell but not [H⁺]. Unfortunately, pH was not measured to evaluate this possibility [43].

Another striking finding of the present experiments is that at exhaustion, muscle lactate concentration and pH vary markedly, not only between subjects, but also comparing interventions and the two legs of a subject. In figures 16 and 17, data on pre- and postexercise muscle lactate at the time of exhaustion are given. The muscle pHs, when determined, reflected the alterations in muscle lactate. It is of note, however, that a given lactate accumulation during EX2 caused a somewhat less pronounced reduction in pH. This finding would indicate a sudden increase in buffer capacity of the muscle or that the extrusion of [H⁺] becomes faster than that for lactate. Buffer capacity of muscle can quickly be altered, for example by an enlarged breakdown of CP. However, this was not increased when comparing EX1 and EX2 (table 1). Muscle temperature also affects pH and it was higher (EX2 + ~0.5 °C), but it is questionable if this can explain the 'damping' of the pH change, as all pH measurements were performed in vitro at 37 °C.

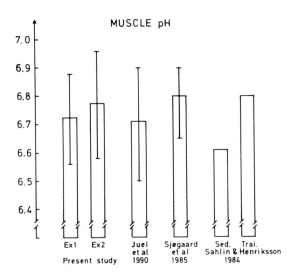

Fig. 17. The mean and range of muscle pH in the present experiments for EX1 and EX2. Included are data from some other studies [from refs 17–19, 23–24].

Taken together this may speak in favor of the explanation for less of an effect on pH to be due to a fast [H⁺] transport. Whatever the explanation may be, it should not distract from the fact that, as in previous experiments on man from our laboratory [17, 19, 21], as well as for an example in studies by Hermansen and Osnes [20] and Sahlin and Henriksson [18], there is no definite value for either muscle lactate or pH when the subjects are exhausted (fig. 17).

With the exercise model used in the present experiment, the site for fatigue is local. The systemic effects of one-legged dynamic knee extensions are minor. Whole body oxygen uptake is less than 1/3 of the subject's maximal oxygen uptake, heart activity is 120–140 beats per minute and catecholamines are low [21]. Although rate of lactate production is high in the active muscle, and its release to the bloodstream is high, elevations in systemic lactate levels are only 5–6 mM. Blood glucose is maintained above 4.5 mM. Point of exhaustion was defined as the inability to maintain the pre-set kicking rate (1 Hz). At that stage, clear indications of fatigue were already present. Force recordings revealed a reduction in amplitude which was compensated for by an increase in duration (fig. 18). Further, a tendency for pushing the leg back to its vertical position was sometimes apparent the

FORCE

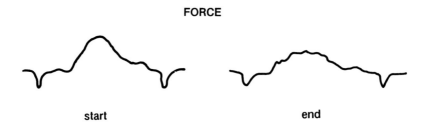

start   end

Fig. 18. Force recordings from 'early' and 'late' during an exhaustive run [example from ref. 23].

closer the exercise came to the point of exhaustion. Thus, definite objective signs of muscle fatigue were present. In this connection, it should be noted that there appears to be no 'transfer' effect of fatigue, i.e. when the exercise was performed with leg 2 after leg 1 had exercised twice, time to exhaustion was nearly identical when comparing EX1 with legs 1 and 2.

The most commonly discussed sites for an action of elevated [H^+] are the inhibition of PFK [47, 48] and a reduced affinity for Ca^{2+}, to bind to troponin [49], or both, in combination. What could be argued is that as ATP is reduced to the same extent in each intervention, the regulation is very tight in controlling and matching ATP utilization with rate of ATP resynthesis. Thus, the elevated [H^+] should retard cross-bridge cycling and usage of ATP, which in turn, will gradually limit muscle tension development and cause exhaustion. At the same time rate of glycolysis is down (note that peak rate of oxygen uptake and thus mitochondrial ATP production is unaltered comparing all interventions used). The question is whether the reduction in pH is large enough to have the proposed effects? In the work on troponin affinity for Ca^{2+}, ST fibers are barely affected with a drop in pH to 6.7, whereas FT fibers are [49]. Thus, the site for fatigue would be the FT fibers, where the drop in ATP resynthesis should also be the most marked. It is questionable, however, whether the reduction in pH, and thus the elevation in [H^+], are large enough to inhibit the PFK activity. The data of Danforth et al. [47] would indicate that this is the case, but more recent exploration of the allosteric regulation of PFK would negate a pH effect until it is reduced to much lower levels than 6.7 [50]. Of note is that with the NMR technique much lower values have been reported for muscle pH at exhaustion [51]. If pH is lowered to 6.0 or even lower values, then pH would be the critical factor causing loss of tension development. We have no indications, however, that other crucial factors are altered. Alteration in ATP (decrease), ADP, AMP and IMP (increase) are similar for all exercise bouts (table 1), and would at first sight appear to cause the same degree of activation, unless the ratio of unbound to bound nucleotides is not altered, which is quite a likely possibility [16, 51]. Elevated cytosolic citrate concentration has been proposed to be an inhibitor of PFK activity. Unfortunately, this was not measured in the present studies, but major differences are unlikely to have occurred, as rate of oxidation was so similar in all experiments, and supply of acetyl-CoA from β-oxidation or de-carboxylation of pyruvate would not have been enhanced in the experiments where rate of glycolysis was reduced the most (experiments IIa, IIb, III).

It should not be excluded that the modification of rate of glycolysis is upstream, i.e. at the level of the activation of phosphorylase b to a. In addition to sympathetic control, AMP, P_i and possible undefined factors related to contractile activity are likely to play a role. This raises an intriguing possibility. In recent studies in man, it has been possible to demonstrate that a drop in contractile force and elongation of half relation time are correlated with a slowing of SR kinetics [52]. This reduction in Ca^{2+} release and uptake with intense contractile activity can occur without any major changes in acidity of the muscle. An alternative scheme of events explaining our finding could then be that the primary site for 'failure' is the SR system, which for still undefined reasons, gradually reduces its rate of releasing and taking up of Ca^{2+}. This will cause less tension development and a slowing of the contraction, which was also observed (fig. 18). What may be gained by the muscle is force generated with slightly less ATP utilization [53, 54].

Another finding to note is that lactate production is lower than reduction in rate of glycolysis. The method used in the present study provides an exact measure of the absolute amount of lactate produced during an exhaustive exercise bout. The explanation for less lactate being accumulated and released from the active muscles cannot be an elevated turnover within the muscle. Instead, the most likely explanation is that the formation of lactate is the net result of rate of pyruvate and NADH production in the cytosol, and their uptakes by the mitochondria. The latter appearing to be unaltered in the present experiments. The smaller production of three carbon skeleton via glycolysis would then result in less lactate being produced. The match between a smaller glycogen depletion and the smaller amount of lactate produced is close.

Another site for failure of the excitation-contracting coupling (E-C) would be the propagation of the action potential (AP) due to ion disturbances over the sarcolemma [55]. During the exhaustive exercise K^+ was released in similar amounts during the different interventions. The absolute value for its loss, as anticipated, being rather small due to the short exercise times. More importantly, however, is the femoral venous K^+ concentration, as this reflects the interstitial K^+ concentration of the muscle when the perfusion is high. The question, however, is whether a venous K^+ concentration of 6 mM is high enough to slow the AP and possibly block its propagation into the T tubuli.

The neural activation of the muscle may also be reduced during an exhaustive exercise bout. A central nervous (cortical) origin is unlikely. A more plausible explanation would be inhibition at the spinal level. Garland and McComas [56] have recently demonstrated reflex inhibition occurring during fatiguing exercise of the soleus muscle. Also during more prolonged exercise in man, a similar mechanism has been proposed based on a reduction in the EMG activity [57]. The question to be raised in relation to the present experiment is: What could activate such a reflex? It is tempting to suggest certain metabolites or ions accumulating in the interstitium of active muscle, triggering sensory receptors of group III or IV nerve fibers. Again, K^+ as well as [H^+] could be the compounds. The scheme which may prevail is that this sensory input causes inhibition of spinal motor nerves. This is likely a gradual phenomenon, and can at first be overcome by the drive from the motor cortex as new motor units are activated. A point is reached, however, where the reflex inhibition causes a lowering of spinal motor activity output, adding to the inability of the muscle to reach the target force and maintain kicking rate. Characteristically, this failure to produce a large enough force recovers very quickly. That is not to say that it is possible to perform for long periods at the same exercise intensity, but after some 20–30 s of rest the exercise can be continued at the same intensity for a brief time [58]. This could be related to the quick normalization of the K^+ homeostasis which occurs upon termination of the exercise. The time course for a change in [K^+] homeostasis is quite different from that of [H^+] (fig. 19), which suggests that [K^+] rather than [H^+] would be the critical factor, or a compound which is as quickly restored to normal level as [K^+].

In conclusion, we have found no coupling between above normal muscle glycogen storage, enhanced glycogenolysis and performance. Prior short exhaustive exercise with only a small glycogen utilization and time for normalization of the acid/base balance of the muscle, cause a significant reduction in performance when the exercise is performed the second time. Glycogenolysis is also reduced and so is lactate production but not the aerobic utilization of pyruvate and its contribution to ATP resynthesis and the alterations in nucleotides and CP/P_i caused by the exhaustive exercise.

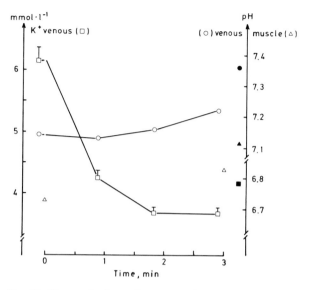

Fig. 19. Changes in femoral venous [K⁺] and pH ([H⁺]) during the first minute of recovery after exhaustive knee-extensor exercise [data from ref. 22]. The filled symbols to the right depict pre-exercise values.

This finding, together with the significant but not very pronounced effect of an accumulation of muscle lactate and lowering of muscle pH prior to start of a second work bout, may indicate that disturbance of the acid/base balance of skeletal muscle is not as crucial as a fatigue factor as is often suggested. Instead, it is pointed at E-C failure coupled with reduced nervous drive possibly due to reflex inhibition on the spinal level as an alternative to explain what causes fatigue and exhaustion in intense short exhaustive exercise. The reduced contractile activation may be linked to a retardation of glycogenolysis without affecting the pyruvate supply for mitochondrial oxidation.

References

1 Christensen EH, Hansen O: Arbeitsfähigkeit und Ernährung. Scand Arch Physiol 1939;81:160–171.
2 Saltin B, Hermansen L: Glycogen stores and prolonged severe exercise; in Blix G (ed): Proc Symp Swedish Nutrition Foun. Uppsala, Almqvist & Wiksell, 1967, vol V, pp 32–46.

3 Bergström J, Hermansen L, Hultman E, Saltin B: Diet, muscle glycogen and physical performance. Acta Physiol Scand 1967;71:140–150.

4 Bergström J, Hultman E, Jorfeldt L, Pernow B, Wahren J: Effect of nicotinic acid on physical working capacity and on metabolism of muscle glycogen in man. J Appl Physiol 1969;26:170–176.

5 Pernow B, Saltin B: Availability of substrates and capacity for prolonged heavy exercise in man. J Appl Physiol 1971;31:416–422.

6 Gollnick PD, Pernow B, Essén B, Jansson E, Saltin B: Availability of glycogen and plasma FFA for substrate utilization in leg muscle of man during exercise. Clin Physiol 1981;1:2742.

7 Richter EA, Galbo H: High glycogen levels enhance glycogen breakdown in isolated contracting skeletal muscle. J Appl Physiol 1986;61:827–831.

8 Spriet LL, Berardinucci L, Marsh DR, Campbell CB, Graham T: Glycogen content has no effect on skeletal muscle glycogenolysis during short-term tetanic stimulation. J Appl Physiol 1990;68:1883–1888.

9 Greenhaff PL, Gleeson M, Maughan RJ: The effects of a glycogen loading regimen on acid-base status and blood lactate concentration before and after a fixed period of high intensity exercise in man. Eur J Appl Physiol 1988;57:254–259.

10 Greenhaff PL, Gleeson M, Maughan RJ: The effects of diet on muscle pH and metabolism during high intensity exercise. Eur J Appl Physiol 1988;57:531–539.

11 Jacobs I: Lactate concentrations after short, maximal exercise at various glycogen levels. Acta Physiol Scand 1981;111:465–469.

12 Klausen K, Sjögaard G: Glycogen stores and lactate accumulation in skeletal muscle of man during intense bicycle exercise. Scand J Sports Sci 1980;2:7–12.

13 Ren JM, Broberg G, Sahlin K, Hultman E: Influence of reduced glycogen level on glycogenolysis during short-term stimulation in man. Acta Physiol Scand 1990;139:467–474.

14 Symons JD, Jacobs I: High-intensity exercise performance is not impaired by low intramuscular glycogen. Med Sci Sports Exerc 1989;21:550–557.

15 Mainwood GW, Renaud JM: The effect of acid-base balance on fatigue of skeletal muscle. Can J Physiol Pharmacol 1985;63:403–416.

16 Porter R, Whelan J: Human Muscle Fatigue: Physiological Mechanisms. London, Pitman Medical, 1981.

17 Juel C, Bangsbo J, Graham T, Saltin B: Lactate and potassium fluxes from skeletal muscle during intense dynamic knee-extensor exercise in man. Acta Physiol Scand 1990;140:147–159.

18 Sahlin K, Henrikson J: Buffer capacity and lactate accumulation in skeletal muscle of trained and untrained men. Acta Physiol Scand 1984;122:331–339.

19 Sjögaard G, Adams RP, Saltin B: Water and ion shifts in skeletal muscle of humans with intense dynamic knee extension. Am J Physiol 1985;248:R190–196.

20 Hermansen L, Osnes JB: Blood and muscle pH after maximal exercise in man. J Appl Physiol 1972;32:304–308.

21 Bangsbo J, Gollnick PD, Graham TE, Juel C, Kiens B, Mizuno M, Saltin B: Anaerobic energy production and O_2 deficit-debt relationship during exhaustive exercise in humans. J Physiol 1990;422:539–559.

22 Saltin B, Kiens B, Savard G: A quantitative approach to the evaluation of skeletal muscle substrate utilization in prolonged exercise; in Benzi G, Packer L, Siliprandi N

(eds): Biochemical Aspects of Physical Exercise. Amsterdam, Elsevier, 1986, pp 235–244.

23 Bangsbo J, Graham TE, Kiens B, Saltin B: Elevated muscle glycogen and anaerobic energy production during exhaustive exercise in man. J Physiol 1991; submitted.

24 Bangsbo J, Graham T, Johansen L, Strange S, Saltin B: Previous exercise and anaerobic energy production during exhaustive exercise in man. J Appl Physiol 1991; submitted.

25 Bangsbo J, Johansen L, Strange S, Saltin B: Muscle metabolism and metabolites fluxes in recovery from exhaustive intense exercise of man; effect of active recovery. J Appl Physiol 1991; submitted.

26 Belcastro AN, Bonen A: Lactic acid removal rates during controlled and uncontrolled recovery exercise. J Appl Physiol 1975;39:932–936.

27 Davies CTM, Knibbs AV, Musgrove J: The rate of lactic acid removal in relation to different baselines of recovery exercise. Int Z angew Physiol 1970:155–161.

28 Dood S, Powers SK, Callender T, Brooks E: Blood lactate disappearance at various intensities of recovery exercise. J Appl Physiol 1984;57:1462–1465.

29 Hermansen L, Stendsvold I: Production and removal of lactate during exercise in man. Acta Physiol Scand 1972;86:191–201.

30 McMaster WC, Stoddard T, Dancan W: Enhancement of blood lactate clearance following maximal swimming. Am J Sports Med 1989;17:472–477.

31 Galbo H, Holst JJ, Christensen NJ: The effect of different diets and of insulin on the hormonal response to prolonged exercise. Acta Physiol Scand 1979;107:19–32.

32 Sherman WM, Costill DL, Fink WJ, Miller JM: Effect of exercise-diet manipulation on muscle glycogen and its subsequent utilization during performance. Int J Sports Med 1981;2:114–118.

33 Asmussen E, Klausen K, Nielsen L, Egelund L, Techow OSA, Tonder PJ: Lactate production and anaerobic work capacity after prolonged exercise. Acta Physiol Scand 1974;90:731–742.

34 Jansson E, Kajser L: Effect of diet on the utilization of blood-borne and intramuscular substrates during exercise in man. Acta Physiol Scand 1982;115:19–30.

35 Grisdale RK, Jacobs I, Cafarelli E: Relative effects of glycogen depletion and previous exercise on muscle force and endurance capacity. J Appl Physiol 1990;69:1276–1282.

36 Boobis L, Williams C, Wotton SA: Influence of sprint training on muscle metabolism during brief maximal exercise in man. J Physiol 1983;342:36–37.

37 Boobis LH: Metabolic aspects of fatigue during sprinting; in Macleod D, Maughan R, Nimmo M, Reilly T, Williams C (eds): Exercise, Benefits, Limits and Adaptations. London, Spon, 1986, pp 116–143.

38 Saltin S, Strange S, Bangsbo J, et al: Central and peripheral cardiovascular responses to electrically induced and voluntary leg exercise. Proc 4th Eur Symp Life Sci Res in Space, Trieste, 1990, pp 591–595.

39 Chassiotis D, Edström L, Sahlin K, Sjöholm H: Activation of glycogen phosphorylase by electrical stimulation of isolated fast-twitch and slow-twitch muscles from rats. Acta Physiol Scand 1985;123:43–47.

40 Chassiotis D: The regulation of glycogen phosphorylase and glycogen breakdown in human skeletal muscle. Acta Physiol Scand 1983;(suppl 518):1–68.

41 Chassiotis D, Hultman E, Sahlin K: Acidotic depression of cyclic AMP accumulation and phosphorylase b to a transformation in skeletal muscle in man. J Physiol 1983;335:197–204.

42 Ren JM, Hultman E: Regulation of phosphorylase a activity in human skeletal muscle. J Appl Physiol 1990;69:919–923.

43 Karlsson J, Bonde-Petersen F, Henriksson J, Knuttgen HG: Effect of previous exercise with arms and legs on metabolism and performance in exhaustive exercise. J Appl Physiol 1975;38:763–767.

44 Klausen K, Knuttgen HG, Forster HV: Effect of preexisting high blood lactate concentration on maximal exercise performance. Scand J Clin Lab Invest 1972;30: 415–419.

45 Benade AJS, Heisler N: Comparison of efflux rates of hydrogen and lactate ions from isolated muscles in vitro. Respir Physiol 1978;32:369–380.

46 Heigenhauser GJF, Lindinger MI, Spriet LL: Lactate proton and ion fluxes in the stimulated perfused rat hindlimb. Clin Physiol 1985;5(suppl 4): 140.

47 Danforth WH: Activation of glycolytic pathway in muscle; in Chance B, Estabrook BW, Williamson JR (eds): Control of Energy Metabolism. New York, Academic Press, 1965, pp 287–297.

48 Ui M: A role of phosphofructokinase in pH dependent regulation of glycolysis. Biochim Biophys Acta 1966;124:310–322.

49 Donaldson SKB: Effect of acidosis on maximum force generation of peeled mammalian skeletal muscle fibres; in Knuttgen HG, Vogel A, Poortmans J (eds): Biochemistry of Exercise. Champaign, Human Kinetics, 1983, pp 123–133.

50 Dobson GP, Yamamoto E, Hochachka PW: Phosphofructokinase control in muscle: Nature and reversal of pH-dependent ATP inhibition. Am J Physiol 1986;250:R71–76.

51 Wilson JR, McCully KK, Mancini DM, Boden B, Chance B: Relationship of muscular fatigue to pH and diprotonated P_i in humans: A ^{31}P-NMP study. J Appl Physiol 1988;64:2333–2339.

52 Gollnick PD, Körge P, Karpakka J, Saltin B: Elongation of skeletal muscle relaxation during exercise is linked to reduced calcium uptake by the sarcoplasmic reticulum in man. Acta Physiol Scand 1991;142:135–136.

53 di Prampero PE, Boutellier U, Marguerat A: Efficiency of work performance and contraction velocity in isotonic tetani of frog sartorius. Pflügers Arch 1988;412:455–461.

54 Kushmerick MJ: Patterns in mammalian muscle energetics. J Exp Biol 1985;115: 165–177.

55 Sjögaard G: Exercise-induced muscle fatigue: the significance of potassium. Acta Physiol Scand 1990;140(suppl 593):1–63.

56 Garland SJ. McComas AJ: Reflex inhibition of human soleus muscle during fatigue. J Physiol 1990;429:17–27.

57 Nicol C, Komi PV, Marconnet P: Fatigue effects of marathon running on neuromuscular performance. Scand J Med Sci Sports 1991;1:10–17.

58 Sahlin K, Ren JM: Relationship of contraction capacity changes during recovery from a fatiguing contraction. J Appl Physiol 1989;67:648–654.

Bengt Saltin, MD, Department of Physiology III, Box 5626, Karolinska Institute, S-114 86 Stockholm (Sweden)

Marconnet P, Komi PV, Saltin B, Sejersted OM (eds): Muscle Fatigue Mechanisms in
Exercise and Training. Med Sport Sci. Basel, Karger, 1992, vol 34, pp 115–130

Increased Metabolic Rate Associated with Muscle Fatigue

Ole M. Sejersted[a, b], Nina K. Vøllestad[a]

[a]Department of Physiology, National Institute of Occupational Health, and
[b]Norwegian University of Sport and Physical Education, Oslo, Norway

Introduction

Reduced performance has usually been coined fatigue [1]. However, reduced performance is preceded by important metabolic changes in the muscle. It is well known that force-generating capacity gradually declines during most kinds of prolonged submaximal muscle activation [2–4]. Several authors call this process fatigue, and it is in this sense that it will be used here. The main scope of this article is to review some recent experiments aimed at identifying the processes that could lead to reduced maximal force. The protocol was chosen so as to avoid lactate accumulation and glycogen depletion.

Eventually, submaximal exercise cannot be sustained at the predefined intensity. At this point it is reasonable to assume that the fatigue has gradually caused the relative force load to rise to an intolerable level. Hence, we postulate that reduced performance or exhaustion occurs when the relative load becomes too high because of reduced maximal force-generating capacity of the muscle.

Protocol and Methods

The subjects were strapped by a seatbelt to a chair and both ankles were connected to a strain gauge as described [4]. The standard exercise protocol was to perform repetitive isometric contractions engaging the quadriceps muscle of both thighs for 6 s. After a 4-second pause contraction was repeated. The load was set to 30% of the maximal voluntary contraction force

(MVC). Before and at intervals during exercise, maximal contractions were performed. Exercise was carried out until exhaustion. Details are given by Vøllestad et al. [4].

Four kinds of experiments will be reported. In one series both surface and intramuscular electromyograms (EMG) were recorded. In addition, the muscle was stimulated directly by electrode pads placed distally and proximally over the quadriceps. Stimuli of 120 V and 100 μs duration were applied as either a train of 8 pulses at 50 Hz (tetanic stimulation) or as single shocks (twitch responses). Stimulation was carried out on the resting muscle before and at intervals during exercise. A force of about 50% of MVC was obtained by tetanic stimulation. Surface EMG from the voluntary contractions was rectified and integrated over 0.5-second periods. Intramuscular EMG obtained by a tungsten electrode slowly advanced through the muscle allowed quantification of firing frequencies of individual motor units (MU) (see Bellamare et al. [5] for details).

The second kind of experiment with the same protocol involved repeated muscle biopsies [4]. The biopsies were analyzed for glycogen, lactate, creatine phosphate (CrP), and ATP. Serial histochemically stained sections allowed fiber typing and quantification of glycogen breakdown in individual fibers [6].

In the third series of experiments femoral arterial blood flow was continuously recorded by an ultrasound Doppler device [7]. Catheters in the femoral artery and vein allowed frequent blood samples to be taken and analyzed for lactate, K^+, and oxygen content [8].

The fourth type of experiment included measurements of intramuscular temperature and heat production rates. Thermocouples of copper and constantan were inserted into the vastus lateralis muscle by means of a Venflon which was withdrawn after insertion of the thermocouple. A reference thermocouple was placed in iced water. The technique has previously been used by Sejersted et al. [9] in kidney tissue and by Edwards et al. [10] in human muscle. The principle is that heat is normally removed entirely by convection from these organs. Hence, during arrest of blood flow convection ceases immediately and the initial temperature rise is proportional to the metabolic rate. Test contractions of 40% MVC lasting for 15 s were sufficient to record the temperature rise accurately. Control experiments with arrested circulation confirmed that convective heat removal was negligible even at lower contraction forces. The rate of heat production was linearly related to relative contraction forces. However, there were large differences between thermocouples, probably due to variable amounts of less metabolically active tissue surrounding the thermocouple (e.g. connective tissue, pools of blood or interstitial tissue fluid).

Mechanical Changes

The target force of 30% MVC was sustained for at least 30 min with duty cycles of 0.6. In this series only a few experiments were continued until exhaustion.

Both MVC and tetanic force at 50 Hz fell linearly and in parallel throughout exercise reaching 55 ± 3% of control after 30 min (fig. 1). Hence, since motor units are recruited throughout exercise, there will be a nonuniform fatigue in different fibers. In some fibers the force reduction must therefore have been much greater than the average for the whole muscle.

Twitch responses of resting muscle were even more reduced (low frequency fatigue) so that peak twitch force was only 38 ± 4% of control after 30 min. Interestingly, contraction time (time from onset of force increment to peak force) became shorter and half relaxation time fell from 60 ± 6 to 37 ± 9 ms (fig. 2). These findings indicate that the muscle became faster which is in clear contrast to other forms of fatigue where lactate accumulation is prominent [11].

Thus, the present protocol induced an easily identifiable form of fatigue which did not interfere with performance until exhaustion occurred. Importantly, the parallel reduction of tetanic and voluntary maximal force rule out any contribution of central fatigue [4].

Motor Control

The IEMG of the MVCs fell slightly initially, but on average remained constant throughout exercise. The IEMG of the 30% contractions, however, increased significantly from 18 ± 1 to 57 ± 15% of those obtained during control MVC after 30 min. This increase may be due to several factors. The most important comprise increased firing rates (rate coding), synchronization and recruitment of new MUs [12].

The intramuscular EMG recording allowed identification of individual MUs. During control MVC firing rates averaged 25 ± 2 Hz. Over the first repeated 30% contractions the mean firing frequency fell from 13 ± 1 to 11 ± 1 Hz. Thereafter, a smaller, but significant rise of about 25% occurred.

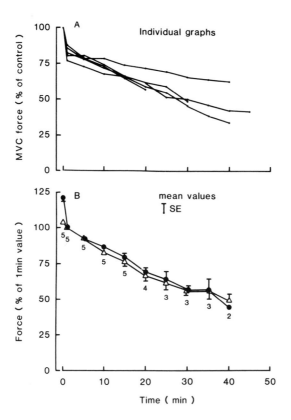

Fig. 1. Changes in force-generating capacity during intermittent isometric contractions at 30% MVC force held for 6 s with 4 s rest between. Upper panel shows changes in MVC force in percent of preexercise MVCs. Lower panel shows mean data (±SE) for MVC force (open triangle) and force response from 50 Hz stimulation (filled circles). Digits denote number of subjects included. (Reproduced from Vøllestad et al. (1988) with permission of the American Physiological Society.)

Taken together these observations indicate that as force declines in the MUs that were recruited initially, new MUs are recruited with no significant decrease in the motor drive of the fatigued fibers. Rather, the data suggest an increased firing rate in the active fibers.

The recruitment pattern might have significant consequences for the metabolic requirement since it has been reported that type II fibers exhibit 5–6 times higher ATP turnover per unit force developed as compared to type I fibers.

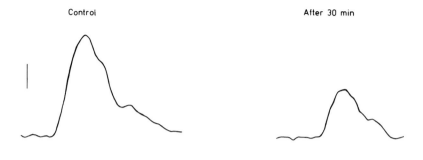

Control   After 30 min

Fig. 2. Typical twitch responses of resting muscle obtained by transcutaneous stimulation of the quadriceps muscle with a single shock of 120 V lasting 100 μs. Control, unfatigued muscle twitch response to the left; twitch response of fatigued muscle after 30 min of voluntary, intermittent, isometric contractions at 30% MVC force to the right. Calibration bars represent 10 *N* and 0.1 s, respectively.

The recruitment pattern was estimated by the glycogen depletion method, using quantitative measurements on single fiber level as described by Vøllestad et al. [6]. Over the first 30 min glycogen depletion was detected in type I fibers and in some type IIA fibers. In the least oxidative fibers, type IIAB and type IIB, no changes in glycogen were observed. However, at exhaustion significant glycogen depletion was observed in all fiber types (fig. 3). These data indicate that type I and some type IIA fibers were activated at an early stage of the exercise, whereas type IIAB and type IIB were recruited as exhaustion was approached.

This recruitment pattern is the same as that seen during dynamic exercise [6]. It closely follows the regular type I → type IIA → type IIAB+IIB seen with increasing force.

Energy Metabolism

The most compelling finding was the almost unchanged level of high energy phosphates, glycogen and lactate for the whole exercise period (fig. 4). Initially CrP fell by 11%, but thereafter remained constant for almost an hour. Glycogen also fell initially, but was paralleled by a fall in protein concentration probably due to fluid uptake by the muscle [4]. No change in ATP could be detected. Lactate rose to 1.5 ± 0.4 mmol · kg wet muscle

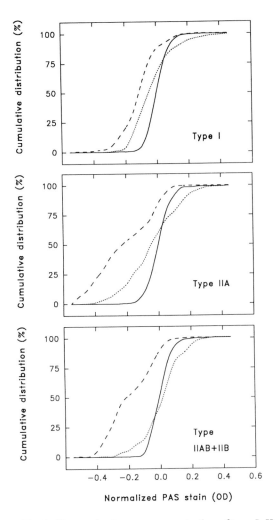

Fig. 3. Changes in glycogen concentration of type I, IIA and IIAB+IIB fibers during intermittent isometric contractions at 30% MVC force. Glycogen depletion pattern is shown by the changes in normalized, cumulative distributions of the optical density of periodic acid Schiff (PAS) stain in single fibers at rest (————), after 30 min exercise (· · · · · ·) and at exhaustion (----).

weight^{-1}, but remained at this level for the first 30 min. Hence, fatigue in the sense of reduced force-generating capacity cannot be due to glycogen depletion, CrP or ATP depletion, or low pH.

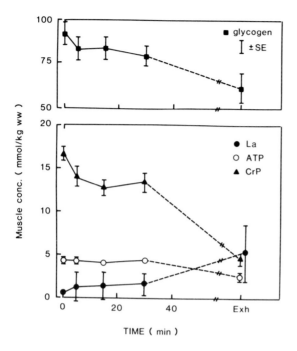

Fig. 4. Glycogen, creatine phosphate (CrP), adenosine triphosphate (ATP) and lactate (La) concentrations in muscle biopsies taken at intervals during intermittent, isometric contractions at 30% MVC force. Mean values ±SE.

At exhaustion very rapid and extensive changes had occurred. CrP had fallen to $27 \pm 5\%$ of control (fig. 4). In one subject fortunate timing of biopsies allowed us to detect a fall of CrP from 15 to 4 mmol · kg wet weight^{-1} over 3 min. ATP fell by 1.5 mmol · kg wet weight^{-1}. Despite these changes in high energy phosphates, lactate accumulation was very modest, reaching 4.8 ± 1.1 mmol · kg wet weight^{-1}, and glycogen fell by about 16 mmol glycosyl units · kg wet weight^{-1} (fig. 4). Our hypothesis is that this abrupt change in metabolite concentrations is the result of a gradually smaller reserve capacity for ATP synthesis compared to ATP demand. Since the anaerobic breakdown was only a few percent of the maximum rate, the demand seems to be on the aerobic metabolism. Hence, experiments were conducted to measure muscle oxygen consumption.

Figure 5 shows how a-fv difference for O_2 and blood flow in the femoral artery gradually increased during exercise. The result in the lower graph shows that at start of exercise leg oxygen consumption increased to 154 ± 18

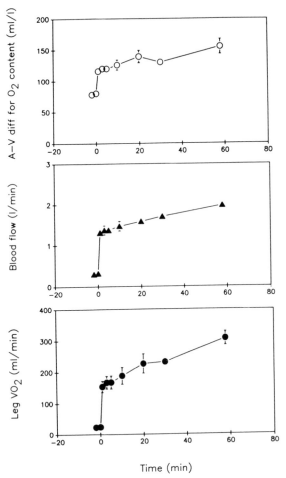

Fig. 5. Arteriovenous difference for blood O_2 content across one leg (upper panel), average femoral arterial blood flow (middle panel) and leg oxygen uptake (lower panel) during and after intermittent, isometric contractions at 30% MVC force. Mean values ± SE.

ml·min^{-1}. Thereafter, a linear increase occurred. By the time of exhaustion, leg oxygen consumption had doubled.

Measurements of muscle release of lactate showed that anaerobic metabolism could contribute to at most 10% of the total energy turnover [8].

To be certain that this increase of metabolism could also be detected locally in the muscle, the heat accumulation technique was used. From figure

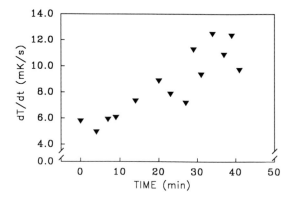

Fig. 6. Metabolic rate, measured as rate of temperature rise (dT/dt) during 15 s test contractions at 40% MVC, repeated every 4–5 min during intermittent, isometric contractions at 30% MVC. Data from one subject [unpubl. results, courtesy E. Saugen].

6 it is clear that the rate of heat accumulation during test contractions at 40% of MVC held for 15 s increased throughout exercise for this thermocouple. On average, heat accumulation rate went up by 65%. Thus, the heat accumulation technique fully confirms the oxygen consumption data. Metabolism in the muscle was approximately doubled over 1 h, whereas performance was kept unchanged at 30% MVC with a duty cycle of 0.6 (fig. 6).

Muscle temperature increased by 2.5 °C over the first 20 min. Thereafter, muscle temperature remained almost constant. Hence, it is clear that the increase of muscle metabolism, at least during the last 30 min of exercise, occurred at almost constant temperature.

It is generally thought that the metabolic rate of exercising muscle increases with a Q_{10} of 3 [13]. However, recent well-controlled studies by Nielsen [14] showed that leg oxygen uptake during exercise at submaximal levels is constant over a temperature range of 3–4 °C. Hence, we conclude that the increase in muscle metabolic rate is most likely not due to any Q_{10} effect. Therefore, energy utilization supported almost entirely by aerobic metabolism is nearly doubled in spite of maintained force output at 30% of control MVC with a duty cycle of 0.6.

Two explanations are possible. The first is based on the evidence in the literature that type II fibers have a higher energy turnover per unit isometric force produced as compared to type I fibers [15, 16]. The second explanation is based on the hypothesis that the relationship between force output and

Table 1. Two possible explanations for the increased energy requirement of muscle at constant isometric force

---

Alternative 1
Constant relationship between force and energy turnover on the individual fiber level.
  Fatigue of type I fibers which are recruited first
  Low energy requirement of fatigued fibers
  Energy requirement of fatigued fibers can probably be met by anaerobic metabolism alone
  Recruitment of new motor units required to maintain overall force
  Fatigue of type II fibers
  Significantly lower force/energy turnover relationship in type II fibers
Consequences for the whole muscle
  Higher ATP turnover at constant force
  Difficult to explain why CrP drops uniformly (in both type I and type II fibers) at exhaustion

Alternative 2
Maintained or increased ATP turnover despite declining force on the individual fiber level
  Fatigue of type I fibers which are recruited first
  Maintained or increased energy requirement in fatigued fibers which are continuously being activated
  Recruitment of new motor units required to maintain overall force
  Fatigue of type II fibers
  Type II fibers may or may not have lower force/energy turnover relationship
Consequences for the whole muscle
  Higher ATP turnover at constant force
  High energy demand can only be met by adequate oxygen supply to all fibers
  CrP drops uniformly in all fibers when energy requirement exceeds O_2 supply

---

ATP is significantly reduced on the cellular level. We only have a rather indirect argument in favor of the latter hypothesis. The arguments are listed in table 1. We cannot exclude that only the changing recruitment pattern contributes to the increased metabolism, but we propose that in addition there is a reduced contraction efficiency at the cellular level. This theory implies that fatigued fibers which generate little force continue to have a high oxygen consumption. The EMG data are compatible with a maintained motor drive to these motor units.

Reduced contraction efficiency can be caused by loose-coupling of either the mitochondrial ATP synthesis or ATP utilization by the contractile apparatus. In favor of the former explanation is the recent finding by Duan et al. [17] that mitochondria in skeletal muscle accumulate Ca^{2+} during strenuous eccentric work.

The latter explanation would require a reduced force generated by each crossbridge attachment with a maintained stoichiometry of ATP hydrolysis per cycle.

Potassium Balance

Potassium has been regarded as a fatigue factor, especially due to the depolarizing effect of increased extracellular concentrations [18]. Potassium concentration increased abruptly in the femoral venous blood when exercise started. It is not possible to tell whether higher transient peaks of K^+ occurred. Measurement with intravasal electrodes indicate that this was not so [Hallén, personal commun.]. Hence, the concentrations we observed were probably close to the highest that were reached, even in the blood stagnant in the muscle during contraction. However, local accumulation of K^+ might well exceed the initial venous peak presently observed. At exhaustion venous K^+ concentration was on average 5.2 ± 0.1 mmol·1^{-1}.

Over the first 10 min the K^+ loss amounted to 3.5 mmol from one leg. This calculation is based on a-v differences and blood flow measurements. Thereafter, the rate of release was reduced, but remained at this level throughout exercise. Figure 7 shows data from one subject. At exhaustion the total muscle loss of K^+ averaged 11 mmol from one quadriceps (approximately 2 kg muscle) [19]. The rise of the plasma concentration is far less than predicted from the accumulated loss provided the distribution volume equalled the total extracellular volume. Hence, K^+ must have been taken up by tissues other than the exercising muscle.

During recovery the negative a-v difference rapidly became positive. Over the first 10 min a reuptake of 1.5 mmol occurred. No significant a-v differences could be observed thereafter. Hence, restitution of the K^+ balance in muscle has an initial rapid component accounting for about 15% of the total loss in this experimental protocol. Restitution of the remaining muscle K^+ deficit must go over a very long period.

The present data give no indication that K^+ affects muscle function by virtue of extracellular accumulation. However, the continuous loss of K^+ is striking and parallels the loss of force-generating capacity.

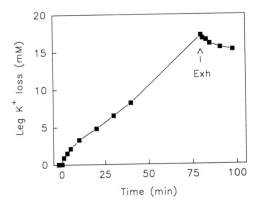

Fig. 7. Accumulated leg K⁺ loss during and after intermittent, isometric contractions at 30% MVC force. Data from one subject calculated from arteriovenous plasma concentration difference across one leg and average plasma flow.

General Discussion and Conclusions

This study shows that the relationship between energy consumption and performance of a muscle may change considerably. Heat production and oxygen consumption may increase significantly at unchanged force output. The mechanism for the altered coupling between metabolism and the contractile proteins remains elusive, but two important factors must be considered: the recruitment pattern of motor units and metabolic changes at the cellular level.

The importance of recruitment was evident from both the EMG data and the glycogen depletion pattern. Integrated surface EMG rose by 170% during exercise, and only a minor part of this increase could be explained by increased firing rates of individual motor units. The simplest explanation is that electrical activity was maintained in fatiguing fibers which developed less and less force, and the EMG signal rose due to recruitment. Glycogen depletion patterns support the EMG data. Additional type I and type IIA fibers were recruited as exercise proceeded. The fast fatiguable fiber types IIAB and IIB were only recruited during the final 30 min of exercise. This recruitment pattern at constant force output fits well with experiments during dynamic exercise [6].

It must be considered that glycogen depletion in the present circumstances did not seem to be closely associated with fatigue. At exhaustion very few fibers were completely empty of glycogen. Hence, it is quite clear that force reduction occurred long before any substrate limitation was present. Even at exhaustion this low intensity exercise was associated with profound decrements in CrP levels when glycogen was still high and lactate remained low. The low stimulation of glycolysis has been commented upon by Vøllestad et al. [4] and is in clear contrast to previous studies with dynamic exercise or electrically stimulated ischemic muscle [20].

The absence of lactate accumulation also precludes low pH as a fatigue factor in these experiments. Hence, force reduction despite maintained stimulation may be due to a defect in the excitation contraction coupling prior to the binding of Ca^{2+} to troponin C. The experiments of Allen et al. [21] and Westerblad et al. [22] may be especially relevant. They showed reduced Ca^{2+} release in isolated muscle cells from *Xenopus laevis* during fatiguing stimulation protocols.

The cause of the increased metabolism could relate both to recruitment pattern and to changes in the individual fibers. There is some evidence from in vitro studies that energy turnover for a given force is much higher in type II as compared with type I fibers [23]. Similar conclusions have been drawn from measurement of heat production rates in human muscles [16]. Recently, Medbø [24] has shown that efficiency during dynamic work is not well related to fiber type composition and work load. Dynamic exercise might not be relevant for the isometric contractions used in the present protocol. Hence, while awaiting further studies, we must conclude that increased recruitment of type II fibers probably contributes to the increasing energy turnover.

On the cellular level the changed coupling between force and energy turnover can occur either with the ATP producing or the ATP consuming processes as discussed by Vøllestad et al. [8] and Vøllestad and Sejersted [25]. Provided the ATP production has become less efficient, energy efficiency in terms of ATP/O_2 ratio would decrease so that all ATP-consuming processes would require a larger O_2 uptake to proceed unaffected. On the other hand, if ATP consumption by myosin ATPase or the sarcoplasmic Ca^{2+}-ATPase has increased, only one or both of these specific processes will be affected. These predictions are presently being investigated.

The continuous, large K^+ loss showed that compensatory reuptake mechanisms were inadequate to maintain the cellular K^+ content. K^+ uptake by the cell is either directly or indirectly driven by the Na,K pump. One main stimulator of pump rate is intracellular Na^+ [26]. It is unlikely that intracellular Na^+ rose continuously, and therefore, cellular gain of Na^+ was probably less pronounced than the loss of K^+ as found by Juel [27] in isolated mouse muscles. Hence, intracellular Na^+ was maintained at the cost of loss of K^+ in fibers that were activated.

Another possible explanation for the continuous loss of K^+ relates to the gradual recruitment of new fibers. When a fiber is activated there is always a rapid initial loss of K^+ until the Na,K pump rate has increased after a few minutes [28]. At the start of exercise many motor units are recruited simultaneously, compatible with a transient and initially rapid K^+ loss. Thereafter, additional fibers that are gradually recruited might go through the same transient rapid K^+ loss, which then becomes evident as a slower, continuous loss of K^+ from the whole muscle. We cannot estimate the relative contribution of this process to the overall K^+ homeostasis of the exercising muscle.

After exercise of high intensity or after sudden decrements in heart rate, K^+ in skeletal muscle or heart muscle is rapidly and completely restored [29]. This was not the case after the present exercise protocol which means that Na,K-pump rate was too slow to drive reuptake of lost K^+, possibly due to low intracellular Na^+ concentration. Since the plasma K^+ concentration was normalized at this point, complete restoration would require movement of K^+ from one tissue to another.

In conclusion, with repetitive isometric contractions, fatigue in the sense of reduced MVC is associated with two major changes in muscle function. First, energy turnover increases gradually, and second, the muscle loses K^+ continually, How these changes are connected through the gradual recruitment of type II fibers and/or the metabolic changes on the cellular level remains to be investigated.

Acknowledgments

We thank Eirik Saugen for providing the heat accumulation data. The study was supported by The Norwegian Research Council for Science and the Humanities and The Royal Norwegian Council for Scientific and Industrial Research.

References

1   Edwards RHT: Human muscle function and fatigue; in Porter R, Whelan J (eds): Human Muscle Fatigue; Physiological Mechanisms. London, Pitman Medical, 1981, pp 1–18.
2   Simonson E, Weiser P: Physiological Aspects and Physiological Correlates of Work Capacity and Fatigue. Springfield, Thomas, 1976.

3 Bigland-Ritchie B, Cafarelli E, Vøllestad NK: Fatigue of submaximal static contractions. Acta Physiol Scand 1986;128:137–148.
4 Vøllestad NK, Sejersted OM, Bahr R, Woods JJ, Bigland-Ritchie B: Motor drive and metabolic responses during repeated submaximal contractions in man. J Appl Physiol 1988;63:1–7.
5 Bellamare F, Woods JJ, Johansson R, Bigland-Ritchie B: Motor-unit discharge rates in maximal voluntary contractions of three human muscles. J Neurophysiol 1983;50:1380–1392.
6 Vøllestad NK, Vaage O, Hermansen L: Muscle glycogen depletion patterns in type I and subgroups of type II fibres during prolonged severe exercise in man. Acta Physiol Scand 1984;122:433–441.
7 Wesche J: The time and magnitude of blood flow changes in the human quadriceps muscles following isometric contraction. J Physiol 1986;377:445–462.
8 Vøllestad NK, Wesche J, Sejersted OM: Gradual increase in leg oxygen uptake during repeated submaximal contractions in humans. J Appl Physiol 1990;68:1150–1156.
9 Sejersted OM, Lie M, Kiil F: Effect of ouabain on metabolic rate in renal cortex and medulla. Am J Physiol 1971;220:1488–1493.
10 Edwards RHT, Hill DK, Jones DA: Heat production and chemical changes during isometric contractions of the human quadriceps muscle. J Physiol 1975;251:303–315.
11 Cady EB, Elshove H, Jones DA, Moll A: The metabolic causes of slow relaxation in fatigued human skeletal muscle. J Physiol 1989;418:327–337.
12 DeLuca CJ: Myoelectrical manifestations of localized muscular fatigue in humans. CRC Crit Rev Biomed Eng 1984;11:251–280.
13 Woledge RC, Curtin NA: Energetic Aspects of Muscle Contraction. London, Academic Press, 1985.
14 Nielsen B: Heat stress causes fatigue! Exercise performance during acute and repeated exposures to hot, dry environments; in Marconnet P (ed): Muscle Fatigue Mechanisms in Exercise and Training. Med Sport Sci. Basel, Karger, 1992, vol 34, pp 208–218.
15 Katz A, Sahlin K, Henriksson J: Muscle ATP turnover rate during isometric contraction in humans. J Appl Physiol 1986;60:1839–1842.
16 Bolstad G, Ersland A: Energy metabolism in different human skeletal muscles during voluntary isometric contractions. Eur J Appl Physiol 1978;38:171–179.
17 Duan C, Delp MD, Hayes DA, Delp PD, Armstrong RB: Rat skeletal muscle mitochondrial [Ca^{2+}] and injury from downhill walking. J Appl Physiol 1990;68:1241–1251.
18 Sjøgaard G: Exercise-induced muscle fatigue: significance of potassium. Acta Physiol Scand 1990;140(suppl 593):1–63.
19 Vøllestad NK, Wesche J, Sejersted OM: Potassium balance of muscle and blood during and after repeated isometric contractions in man (abstract). J Physiol 1991; 483:202 P.
20 Hultman E, Spriet LL: Skeletal muscle metabolism, contraction force and glycogen utilization during prolonged electrical stimulation in humans. J Physiol 1986;374:493–501.
21 Allen DG, Lee JA, Westerblad H: Intracellular calcium and tension during fatigue in isolated single muscle fibres from Xenopus laevis. J Physiol 1989;415:433–458.

22 Westerblad H, Lee JA, Lamb AG, Bolsover SR, Allen DG: Spatial gradients of intracellular calcium in skeletal muscle during fatigue. Pflügers Arch 1990;415:734–740.

23 Curtin NA, Woledge R: Energy changes and muscle contractions. Physiol Rev 1978;58:690–761.

24 Medbø JI: Type I and type II fibres work with the same mechanical efficiency during bicycling; in Marechal G, Carraro U (eds): Muscle and Motility, vol 2. Proc XIXth Eur Conf in Brussels. Andover, Hampshire, Intercept Ltd, 1990, pp 303–308.

25 Vøllestad NK, Sejersted OM: Biochemical correlates of fatigue. A brief review. Eur J Appl Physiol 1988;57:336–347.

26 Sejersted OM, Wasserstrom JA, Fozzard HA: Na,K pump stimulation by intracellular Na in isolated, intact sheep cardiac Purkinje fibers. J Gen Physiol 1988;91:445–466.

27 Juel C: Potassium and sodium shifts during in vitro isometric muscle contraction, and the time course of the ion-gradient recovery. Pflügers Arch 1986;406:458–463.

28 Sejersted OM, Hallen J: Na,K homeostasis of skeletal muscle during activation; in Muscular Function in Exercise and Training. Med Sport Sci, vol 26. Basel, Karger, 1987, pp 1–11.

29 Sejersted OM: Maintenance of Na,K-homeostasis by Na,K-pumps in striated muscle; in Skou JC, Nørby JG, Maunsbach AB, Esmann M (eds): The Na$^+$,K$^+$-pump. B. Cellular Aspects. Prog Clin Biol Res, vol 268B. New York, Liss, 1988, pp 195–206.

Ole M. Sejersted, MD, Institute for Experimental Medical Research,
Ullevaal Hospital, N–0407 Oslo 4 (Norway)

Marconnet P, Komi PV, Saltin B, Sejersted OM (eds): Muscle Fatigue Mechanisms in
Exercise and Training. Med Sport Sci. Basel, Karger, 1992, vol 34, pp 131–139

Damage to Skeletal Muscle during Exercise:
Relative Roles of Free Radicals and
Other Processes

M.J. Jackson[1]

Muscle Research Centre, Department of Medicine, University of Liverpool, UK

Muscle pain, together with some loss of force generation, is a common
phenomenon following strenuous or unaccustomed exercise in man [1]. On
closer examination these changes may be associated with morphological
and ultrastructural disruption of muscle architecture and with the appearance of biochemical markers of muscle damage such as the release of
intracellular components (e.g. muscle cytosolic enzymes) into the bloodstream.

There are two types of muscle pain which can be demonstrated to occur
following different forms of muscle exercise, one occurring during or
immediately following high-intensity exercise, the other with delayed onset
developing more than 24 h after exercise. This latter form of pain appears to
be provoked by eccentric use of the muscle, while immediate onset pain
appears to be initiated by concentric contractions of muscles [2].

Severe damage to muscle is associated with an increased flux of
cytoplasmic components of muscle into the bloodstream and assay of such
components has proven to be very useful as a biochemical marker of muscle
damage. Assay of creatine kinase activity is perhaps the most widely used of
such markers, but assays of other intramuscular components (e.g. aldolase
activity, myoglobin concentration or carbonic anhydrase III concentration)
have been shown to be useful as markers of damage. Creatine kinase (CK)
has the particular advantage over other potential indicators that it exists in
isoforms with different tissue specificities which can be assayed relatively

[1] The author would like to thank the many coworkers who have contributed to
various aspects of the work described and the Muscular Dystrophy Group of Great
Britain and Northern Ireland for continuing financial support.

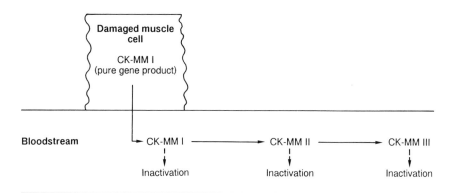

Fig. 1. Schematic representation of the release of CK from muscle and the appearance of CK-MM isoforms in the circulation. The kinetics of the proteolytic modification of CK-MM I and CK-MM II isoforms appear to be predictable and if known will allow a prediction of the elapsed time since an episode of muscle damage from the relative proportions of the various isoforms present in the circulation.

easily, thus CK-MM is the major form of the enzyme found in skeletal muscle whereas CK-MB and CK-BB are of primarily cardiac and brain origin, respectively. A further recent refinement of the assay of creatine kinase has utilised the recognition that CK-MM released into human plasma is proteolytically modified by carboxypeptidases to produce two further isoforms retaining creatine kinase activity [3]. These three isoforms have been designated CK-MM I, II and III, and appear to be released into the bloodstream and modified as shown in figure 1. The conversion of CK-MM I to CK-MM II etc., appears to occur by entirely predictable kinetics and hence analysis of the total CK activity together with the proportions of the individual CK-MM isoforms can provide information not only on the amount of muscle damage which has occurred, but also the time which has elapsed since the episode of muscle damage [3].

This technique has not been frequently applied to studies of muscle damage during exercise, but is an area worthy of further study. In particular, measurement of CK-MM isoforms following commonly undertaken exhaustive exercise such as marathon running, in combination with biochemical studies of substrate supply to muscles, would allow retrospective interpretation of the intramuscular metabolic changes which occur at around the time of onset of muscle damage.

The activity of CK in serum or plasma peaks at 24–48 h after prolonged exercise and 4–7 days after experimental eccentric exercise [4]. The extended time course of the latter situation has prompted many investigations of the potential mechanisms involved [2], but these have not yet been defined. It is relevant in terms of differentiating between the involvement of concentric and eccentric exercise in the genesis of muscle damage following common sports exercise to compare the time course of elevations of plasma creatine kinase activity described above to those seen in athletes post-competition. In most athletes muscle damage is minor and transitory with large elevations of circulatory CK activity only being seen after marathon running or competitions over longer distances. The elevated CK activity in such subjects peaks at 24–48 h indicating that the cause is not likely to be related to eccentric use of the muscles [5].

Involvement of Free Radicals in Exercise-Induced Muscle Damage

A role for oxygen in the pathogenesis of muscle damage during exercise is an attractive possibility because the generation of available energy supplies for the muscle by the processes of mitochondrial respiration depend on a vastly increased oxygen throughput during exercise. The reduction of oxygen to water which normally occurs is thought to generate a certain amount of the one electron reduction product, the superoxide ion (O_2^-), which has been implicated in the initiation of damage to tissues in many situations [6]. Superoxide radicals themselves do not appear to be particularly damaging to biomolecules, but much evidence suggests that they can be readily converted to the hydroxyl radical (OH) which is capable of interacting with, and degrading, DNA, protein or lipids [7].

Many workers have attempted to examine the possibility that radical species may be involved in the damage to muscle which accompanies exercise, but (in common with studies of the role of radicals in disease states) firm acceptance or rejection of these hypotheses has not been possible because of a lack of suitable analytical techniques to study free radicals in biological tissues [8]. Several early workers in this field described elevated levels of non-specific indicators of free radical-mediated lipid peroxidation in rats [9, 10] and man [11] following substantial exercise, and Dillard et al. [11] also demonstrated an apparent protective effect of vitamin E supplementation on this process.

Since the major role of vitamin E in the body has been proposed to be as a lipid soluble antioxidant, several workers have now examined the possibility that manipulation of body vitamin E content would influence the resistance of the animal or man to exercise-induced muscle damage. In particular, Quintanilla and Packer [12] have studied the effects of exercise on vitamin E-deficient animals and human volunteers supplemented with vitamin E and other antioxidants [13] and concluded that the antioxidant capacity of muscle influences its ability to withstand exercise-induced damage. We have also attempted to examine the effect of manipulation of tissue vitamin E content on the ability of muscles to withstand damage induced by excessive contractile activity both in vitro and in vivo [14] and have found the vitamin E content to influence this process. However, further work by our group has cast doubt on our original conclusion and has shown that the vitamin E content of muscle can influence its response to damage caused by non-free radical processes [15, 16].

Recently, a number of workers have attempted to examine various indicators of potential free radical reactions in athletes following exercise. Duthie et al. [17] have reported that the tocopherol content of erythrocytes increased post-exercise, but that paradoxically, erythrocyte susceptibility to in vitro lipid peroxidation was markedly elevated. The same group have also reported that training has a similar effect on erythrocyte tocopherol content and the activity of various erythrocyte antioxidant enzymes [18]. These results are taken to indicate an adaptive increase in antioxidant capacity of subjects to help withstand the increased oxidative stress of exercise, although the design of the experiments does not exclude dietary modification as a possible cause of these changes.

In alternative studies designed to directly examine the possibility that free radicals are involved in the mechanisms of exercise-induced damage to muscle both Davies et al. [19] and our group [20] have used electron spin resonance (ESR) to try and directly detect radicals in muscle.

Examples of the type of ESR spectra obtained from animal muscle are shown in figure 2. These show the enhancing effect of repetitive electrical stimulation on the ESR signal of g-value 2.004. The nature of this signal detected in all muscles examined is not clear, although it has been suggested to be mainly derived from a semiquinone type of molecule [20].

Both of these ESR studies therefore support the overall hypothesis of increased free radical activity during exercise, but further data we have obtained suggest that the ESR signals seen may not necessarily derive from the form of paramagnetic material to which they were originally ascribed and the original conclusions are therefore open to question [21].

It is obvious that much further work is required in this area in order to fully evaluate the role of free radicals in exercise-induced muscle damage. The lack of reliable techniques for detecting increased free radical activity in tissues, together with the inherent problems involved in attempting to standardize and control intervention studies with antioxidant nutrients in

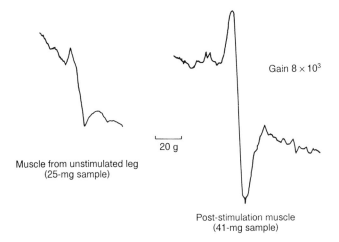

Gain 8×10^3

20 g

Muscle from unstimulated leg
(25-mg sample)

Post-stimulation muscle
(41-mg sample)

Fig. 2. Electron spin resonance signals (g = 2.004) from rat gastrocnemius muscles with and without 30 min electrical stimulation inducing excessive contractile activity. Redrawn from Jackson et al. [20].

animals or man subjected to different exercise protocols has caused considerable confusion in this area. It is hoped that future studies will clarify the role of free radicals in exercise-induced muscle damage and recent studies in which novel chemical compounds were used to attempt to examine the effect of radical scavengers on muscle function [22] appear to indicate one way in which our knowledge and understanding of these processes may be increased.

Alternative Mechanisms by which Muscle May Be
Damaged during Exercise

(a) Mechanical trauma is one possible alternative way by which muscle may be damaged during exercise although very little work has been undertaken to examine this area.

(b) Depletion of muscle energy stores is known to occur in isolated muscles subjected to hypoxic contractile activity [23] and has been suggested to occur in individual muscle fibres of certain animal species following substantial exercise [24]. Little work in this area has been undertaken in man, but it appears to be worthy of further investigation.

(c) A failure of intracellular calcium homeostasis has been implicated in muscle damage in various pathological disorders of muscle [8] and results

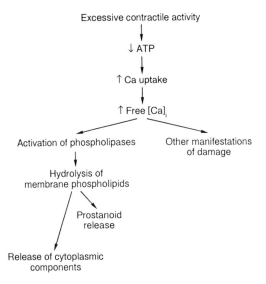

Fig. 3. Schematic representation of a possible mechanism by which excessive contractile activity may induce cytosolic enzyme release from isolated muscles.

from excessive contractile activity of muscle in vitro [25]. Some evidence has been presented that excessive exercise in vivo leads to a failure of muscle calcium homeostasis [1, 26]. A significant sustained elevation of intracellular calcium is toxic to the muscle cell in a number of ways. Calcium activation of proteolytic enzymes [27], overload of mitochondria [28] and activation of phospholipase enzymes [29] have all been implicated in the processes of muscle cell degeneration.

Much of the current work in our laboratory has concentrated on mechanisms (b) and (c) as possible routes by which muscle damage can occur. In the author's opinion there is now considerable evidence in support of a key role for a failure of muscle calcium homeostasis in various features of contractile activity-induced damage. An outline of the manner in which excessive contractile activity might cause efflux of intracellular enzymes via a failure of muscle calcium homeostasis is shown in figure 3, while a summary of our evidence that these processes can occur in muscle is shown in figure 4. It is important to note that this is a possible mechanism for muscle *damage* induced by contractile activity not muscle *fatigue* and that such a system would only become activated following very severe contractile activity.

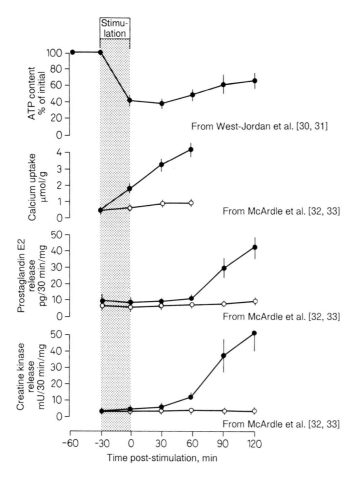

Fig. 4. Effect of repetitive electrically stimulated tetani (for 0.5 see every 2 s) on ATP content, calcium uptake, prostaglandin E_2 release and creatine kinase release from isolated rodent muscles. ● = Stimulated muscles; ○ = non-stimulated control muscles.

Conclusions

It is clear that despite a number of studies there is no consensus concerning the nature of the mechanisms underlying damage to skeletal muscle during exercise and of the relative importance of free radicals in these processes. Several substances have now been reported to reduce exercise-induced muscle damage in experimental situations, but most of the models used are of a specialized nature and/or the results of such debatable relevance that trials in athletes are not warranted. However, it is obvious from a number of studies that training in the appropriate exercise is an excellent way of avoiding excessive muscle damage during or following later exposure to that exercise protocol and as such is probably the best way of avoiding substantial muscle damage during exercise.

References

1 Amelink GJ: Exercise induced muscle damage; PhD thesis, University of Utrecht, 1990.
2 Jones DA, Newham DJ, Round JM, Tolfree SEJ: Experimental human muscle damage: Morphological changes in relation to other indices of damage. J Physiol 1986;375:435–448.
3 Page S, Jackson MJ, Coakley J, Edwards RHT: Isoforms of creatine kinase-MM in the study of skeletal muscle damage. Eur J Clin Invest 1989;19:185–191.
4 Newham DJ, Jones DA, Edwards RHT: Plasma creatine kinase changes after eccentric and concentric contractions. Muscle Nerve 1986;9:59–63.
5 Bar PR, Amelink GJ, Jackson MJ, Jones DA, Bast A: Aspects of exercise induced muscle damage; in Proc FIMS World Congr Sports Medicine, in press.
6 McCord JM: Oxygen-derived free radicals in post-ischaemic tissue injury. N Engl J Med 1985;312:159–163.
7 Halliwell B, Gutteridge JMC: Free Radicals in Biology and Medicine. Oxford, Clarendon, 1985.
8 Jackson MJ, Edwards RHT: Free radicals, muscle damage and muscular dystrophy; in Quintanilla A (ed): Reactive Oxygen Species in Chemistry, Biology and Medicine. New York, Plenum, 1988, pp 197–210.
9 Brady PS, Brady LJ, Ullrey DE: Selenium, vitamin E and the response to swimming stress in the rat. J Nutr 1979;109:1103–1109.
10 Gee DL, Tappel AL: The effect of exhaustive exercise on expired pentane as a measure of 'in vivo' lipid peroxidation in the rat. Life Sci 1981;28:2425–2429.
11 Dillard CJ, Litov RE, Savin WM, Tappel AL: Effects of exercise, vitamin E and ozone on pulmonary function and lipid peroxidation. J Appl Physiol 1978;45:927–932.
12 Quintanilla AT, Packer L: Vitamin E, physical exercise and tissue oxidative damage; in Porter R, Whelan J (eds): Ciba Fdn Symp Ser No 101: Biology of Vitamin E. London, Pitman, 1983, pp 56–69.
13 Packer L, Vigue C: Human exercise: Oxidative stress and antioxidant therapy; in Benzi G (ed): Advances in Myochemistry, vol 2. London, Libbey, 1989, pp 1–20.
14 Jackson MJ, Jones DA, Edwards RHT: Vitamin E and skeletal muscle; in Porter R, Whelan J (eds): Ciba Fdn Symp Ser No 101: Biology of Vitamin E. London, Pitman, 1983, pp 224–239.
15 Phoenix J, Edwards RHT, Jackson MJ: Inhibition of calcium-induced cytosolic enzyme efflux from skeletal muscle by vitamin E and related compounds. Biochem J 1989;287:207–213.

16 Phoenix J, Edwards RHT, Jackson MJ: Effects of calcium ionophore on vitamin E deficient muscle. Br J Nutr 1990;64:245–256.

17 Duthie GG, Robertson JD, Maughan RJ, Morrice PC: Blood antioxidant status and erythrocyte lipid peroxidation following distance running. Arch Biochem Biophys 1990;282:78–83.

18 Robertson JD, Maughan RJ, Duthie GG, Morrice PC: Increased blood antioxidant systems of runners in response to training load. Clin Sci 1991;80:611–618.

19 Davies KJA, Quintanilla AT, Brooks GA, Packer L: Free radicals and tissue damage produced by exercise. Biochem Biophys Res Commun 1982;107:1198–1205.

20 Jackson MJ, Edwards RHT, Symons MCR: Electron spin resonance studies of intact mammalian skeletal muscle. Biochim Biophys Acta 1985;847:185–190.

21 Johnson K, Sutcliffe L, Edwards RHT, Jackson MJ: Calcium ionophore enhances the electron spin resonance signal from isolated skeletal muscle. Biochim Biophys Acta 1988;964:285–288.

22 Novelli GP, Bracciotti G, Fulsin S: Spin trappers and vitamin E prolong endurance to muscle fatigue in mice. Free Rad Biol Med 1990;8:9–13.

23 Jones DA, Jackson MJ, Edwards RHT: The release of intracellular enzymes from an isolated mammalian skeletal muscle preparation. Clin Sci 1983;65:193–201.

24 Foster CVL, Harman J, Harris RC, Snow DH: ATP distribution in single muscle fibres before and after maximal exercise in the thoroughbred horse. J Physiol 1986; 378:64P.

25 Jones DA, Jackson MJ, McPhail G, Edwards RHT: Experimental muscle damage: The importance of extracellular calcium. Clin Sci 1984;66:317–322.

26 Gollnick PD, Hodgson DR, Byrd SK: Exercise-induced muscle damage: a possible link to failures in calcium regulation; in Benzi G (ed): Advances in Myochemistry, vol 2. London, Libbey, 1989, pp 339–350.

27 Ebashi S, Sugita H: The role of calcium in physiological and pathological processes of skeletal muscle; in Aguayo AJ, Karpati G (eds): Current Topics in Nerve and Muscle Research. Amsterdam, Excerpta Medica, 1979, pp 73–84.

28 Wrogemann K, Pena SJG: Mitochondrial overload: A general mechanism for cell necrosis in muscle diseases. Lancet 1976;ii:672–674.

29 Jackson MJ, Jones DA, Edwards RHT: Experimental skeletal muscle damage: The nature of the calcium-activated degenerative processes. Eur J Clin Invest 1984;14: 369–374.

30 West-Jordan JA, Martin PA, Abraham RJ, Edwards RHT, Jackson MJ: Energy dependence of cytosolic enzyme efflux from rat skeletal muscle. Clinica Chim Acta 1990;189:163–172.

31 West-Jordan JA, Martin PA, Abraham RJ, Edwards RHT, Jackson MJ: Energy metabolism during damaging contractile activity in isolated skeletal muscle. A ^{31}P-NMR study. Clinica Chim Acta (in press).

32 McArdle A, Edwards RHT, Jackson MJ: Effects of contractile activity on muscle damage in the dystrophin-deficient mdx mouse. Clin Sci 1991;80:367–371.

33 McArdle A, Edwards RHT, Jackson MJ: ^{45}Calcium accumulation by isolated muscles from dystrophin-deficient mdx mice. J Physiol 1991;434:62P.

M.J. Jackson, PhD, Muscle Research Centre, Department of Medicine,
University of Liverpool, PO Box 147, GB–Liverpool L69 3BX (UK)

Marconnet P, Komi PV, Saltin B, Sejersted OM (eds): Muscle Fatigue Mechanisms in
Exercise and Training. Med Sport Sci. Basel, Karger, 1992, vol 34, pp 140–161

Impaired Lactate Exchange and Removal Abilities after Supramaximal Exercise in Humans

S. Oyono-Enguelle[a], *H. Freund*[b], *J. Lonsdorfer*[a], *A. Pape*[c]

[a]Laboratoire de Physiologie Appliquée, Faculté de Médecine, Strasbourg;
[b]Laboratoire de Pharmacologie Cellulaire et Moléculaire, Faculté de Pharmacie,
Illkirch-Graffenstaden; [c]Centre de Recherches Nucléaires,
IN2P3-CNRS/Université Louis Pasteur, Strasbourg, France

Introduction

Lactate in blood and muscle has been extensively investigated during and after maximal or supramaximal exercise in humans. These studies aimed principally at clarifying the fate of lactate after exercise [Margaria et al., 1933; Margaria and Edwards, 1934; Hermansen and Vaage, 1977; Astrand et al., 1986; Peters-Futre et al., 1987], the relationships between lactate and acid base status and electrolyte balance [Sahlin et al., 1976; Sejersted et al., 1982, 1984], the relationships between high-energy phosphate and lactate [Harris et al., 1977; Karlsson and Saltin, 1970; Karlsson 1971], and the bioenergetic aspects of lactate production [Di Prampero, 1981]. However, very few studies on lactate kinetics during and after maximal or supramaximal exercise are reported in the literature. The main difficulty is the application of methods using labeled or unlabeled lactate during and after maximal and supramaximal exercise. Such exercise is of short duration and generates large transitory variations in blood lactate concentration. Infusion techniques require administration of tracer until a blood lactate steady state is reached [Connor and Woods, 1982; Heteneyi et al., 1983; Freminet and Minaire, 1984]. This requires a lapse of time larger than the time for which the exercise can be sustained even if this delay may be shortened by using a priming dose [Connor and Woods, 1982; Heteneyi et al., 1983]. In the single injection technique, several samples have to be taken to estimate accurately the decay of tracer. These requirements, in general, are not fulfilled in exercise of short duration and of very heavy intensity. As for the Steele method enabling calculations of metabolic fluxes in systems not in the steady state, according to Stanley et al. [1985], it supplies erroneous data when used during near maximal exercise. Furthermore, in all these injection or infusion techniques, the lactate space is considered to be a single compartment. However, this mono-compartment representation cannot account for lactate concentration differences observed during and after exercise, namely between active muscles and arterial blood [Diamant et al., 1968; Karlsson and Saltin, 1970; Karlsson, 1971], arterial and femoral venous blood [Hermansen et al., 1975; Hermansen and Vaage, 1977; Freund and Zouloumian, 1981b], arterial and brachial venous blood [De Coster et al., 1968; Olsen and Strange-Petersen, 1973; Poortmans et al., 1978; Freund and Zouloumian, 1981b; Oyono-Enguelle et al., 1989], the differences giving evidence that under such conditions, the lactate distribution is not homogenous. Theoretically, for an accurate study of lactate movements in the body, the mass conservation law [Zouloumian and Freund, 1981a] should be applied to each site composing the lactate space. But such an application would present many difficulties (fig. 1).

Several other points are also of importance in lactate kinetics studies: (i) Norwich and Wassermann [1986] have recently questioned the validity of metabolic flux calculations when the sources of the tracer and tracee are different; (ii) controversy still persists about the site of injection and of sampling [Okajima et al., 1981; Stanley et al., 1985]; (iii) sodium lactate injection or perfusion modifies plasma volume, osmolarity, and acid base status, so that the generated situation is far from a physiological one [Connor and Woods, 1982; Layzer et al., 1969; Cohen, 1979]; (iv) based on ethical considerations the use of labeled lactate in humans is not authorized in all countries [Freminet and Minaire, 1984]; (v) interpretation of information obtained on lactate metabolism with labeled lactate faces several problems, especially recycling ones, as reviewed by Freminet and Minaire [1984]. In order to overcome some of these difficulties Freund and co-workers have introduced and developed a methodology allowing lactate kinetic studies after muscular exercise in humans [Freund and Gendry, 1978; Freund and Zouloumian, 1981a, b; Zouloumian and Freund, 1981a, b]. This method depends on the following assumptions:

The natural lactate accumulated in the organism at the end of exercise can be used as its own tracer. It can be pointed out that such a lactate load differs from injected or infused lactate in only two closely associated points: the lactate produced in muscles being its own tracer, the source of tracer and

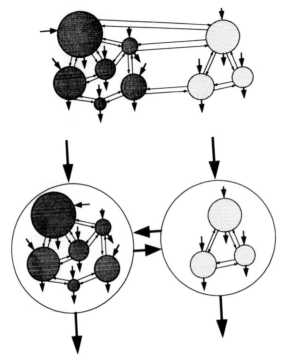

Fig. 1. Top: Schematic diagram of a rigorous application of the mass conservation law to several sites of the lactate space. Bottom: Differentiation of the lactate space into two compartments, the previously worked muscles (M) and the remainder of the space where lactate is distributed (S). For further details, refer to the text.

tracee is the same. Thus, this new methodology is not concerned by the objection put forward by Norwich and Wasserman [1986].

Blood lactate curves supply information on the lactate distribution and displacements in the body. Indeed, as the different sites of the lactate space are irrigated by blood, displacements of lactate in or between these sites should be reflected in blood lactate variations.

Lactate movements in the body can be studied from arterial blood, which has similar lactate concentration as mixed venous blood [Mitchell and Cournand, 1955; Holmgren, 1959; Harris et al., 1965].

The information on lactate movements and distribution becomes available through an appropriate mathematical analysis of blood lactate time courses.

A mathematical model of the type:

$$La(t) = La(0) + A_1(1-e^{-\gamma_1 \cdot t}) + A_2(1-e^{-\gamma_2 \cdot t}), \tag{1}$$

could be fitted accurately to the arterial lactate recovery curves from muscular bicycle exercise whatever its type, intensity and duration. In equation 1, $La(t)$ and $La(0)$ are the arterial lactate concentrations at time t and at the beginning of the recovery, A_1, A_2, γ_1, γ_2 are, respectively, the amplitudes and the velocity constants of the two exponential functions.

A two-compartmental model of lactate distribution in the body appears the simplest but nevertheless realistic explanation of this particular evolution [Freund and Zouloumian, 1981a, b; Zouloumian and Freund, 1981a]. In this model (M) represents the previously worked muscles, and (S) the remainder of the space where lactate is distributed (fig. 1). This methodology has been extensively used by Freund and Gendry [1978], Freund and Zouloumian [1981a, b], Freund et al. [1984, 1986, 1989, 1990], Oyono-Enguelle et al. [1989, 1990].

The aim of the present work is to study the particular blood lactate evolution pattern after exhaustive exercise which seems not to have been previously mentioned, and to try to clarify some of the underlying mechanisms. The experimental results already collected for 4 subjects [Freund et al., 1986] will be used again. At the time, it had already been mentioned that the lactate curves displayed a particular evolution pattern after supramaximal exercise. But since the objective of this earlier study was to assign a physiological meaning to the coefficients of equation 1 as well as to assess their dependence on absolute work rate, the specific aspects of lactate kinetics after supramaximal exercise had not been pursued.

Methods

Subjects and Experimental Protocol
The anthropometric and physiological characteristics of the subjects are listed in table 1. Three (S1, S2, and S3) were untrained students, while one (S4) was a trained subject, regularly involved in physical activities. Details of the experimental procedures and an explanation of the risks involved were provided to the subjects before obtaining their written consent to participate in the experiments. Subjects were tested in the morning at the normal room temperature of 21–23 °C.after a light standard breakfast taken about 2 h before. Heart rate was monitored throughout the experiment from a continuous electrocardiogram as a safety and check procedure. The exercise was performed in a supine position on a mechanically braked cycle ergometer (Fleisch ergometer) at a constant pedaling frequency of 60 rpm. The experiments involved in order, a 10-min

Table 1. Physical and anthropometric characteristics of the subjects

Subject	Age years	Body mass kg	Height cm	PWC170 W	$\dot{V}_{0_2}$ max		Maximal aerobic power, W
					$ml \cdot min^{-1}$	$ml \cdot min^{-1} \cdot kg^{-1}$	
S1	21	61.0	175	200	3,220	52.78	270
S2	22	63.1	178	182	3,400	53.88	280
S3	27	64.0	178	170	3,500	54.68	270
S4	25	66.9	176	255	4,100	61.28	300

PWC170 = Physical work capacity at 170 beats $\cdot$ min^{-1}

rest, a first 3 min 0 W work rate exercise, a 10-min recovery, a second 3 min moderate exercise (1.95 ± 0.4 W·kg^{-1}), a 60-min recovery and the supramaximal exercise at 107–115% $\dot{V}_{O2}$ max (4.53–5.08 W·kg^{-1}) followed by a final recovery of about 120 min. Only the supramaximal exercise and the subsequent recovery will be considered in this study.

Blood Sampling and Analysis
An indwelling 1-mm-diameter polyethylene catheter was placed through a Cournand needle into the brachial artery under local anesthesia. After heparinization of the subjects (heparin, 100 IU·kg^{-1}) to prevent blood clotting, the catheter was connected to an automatic analyzer comprised of a peristaltic pump (Ismatec), a dialyzer (Technicon), an oil bath at 40 °C (Technicon), a detector (LKB 2138 Uvicord S; operating wavelength: 365 nm), and a strip-chart graphic recorder (Hewlett-Packard). Blood was drawn continuously at a rate of 0.32 ml·min^{-1} throughout the experiment. Lactate concentrations were measured by an enzymatic method according to the flow diagram described by Freund et al. [1970] which is a modification of previously described methods [Minaire et al., 1965; Freund, 1967]. Concentration data were delivered at 30-second intervals.

Mathematical Analysis
The individual arterial lactate recovery curves were fitted to equation 1 by means of an iterative nonlinear regression technique. This equation was applied to lactate recovery curves from supramaximal exercise as was done previously for submaximal and maximal exercise [Freund and Zouloumian, 1981a, Freund et al., 1984] in order to avoid introducing any arbitrary distinction between submaximal-maximal and supramaximal exercise.

Results

For the healthy untrained subjects of the present study (S1, S2, and S3), after 3 min supramaximal exercise at 107–115% $\dot{V}_{0_2}$ max, the arterial lactate recovery curve was observed to exhibit a transitory flattening of the lactate

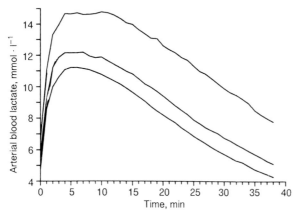

Fig. 2. Arterial lactate recovery curves from 3-min supramaximal exercise (107–115% $\dot{V}_{0_2}$ max, 4.53–5.08 W·kg⁻¹) in untrained subjects. From top to bottom, respectively, the lactate recovery curves of S1, S2, and S3. The upper two show a genuine plateau, while that of S3 has also been considered to level off from min 4 to min 9.

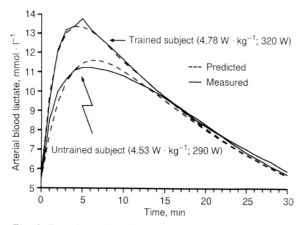

Fig. 3. Experimental and predicted arterial lactate recovery curves from 3-min supramaximal exercise at 107% $\dot{V}_{0_2}$ max. The curve for the trained subject (S4) does not show a plateau; the fit of equation 1 is good. The fit of equation 1 to the arterial lactate recovery curve of an untrained subject (S3) is less good.

concentration curve between the well established increasing and decreasing phases. A clearly identifiable lactate plateau maximum lasting for about 6 min before declining is seen in the lactate recovery curves of two untrained subjects (S1 and S2), Even for S3 who exercised at 107% $\dot{V}_{0_2}$ max (4.53 W·kg⁻¹), from min 4 to min 9 of the recovery, the variations in lactate

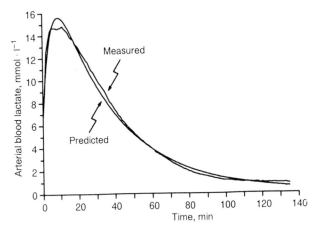

Fig. 4. Example of a fit of equation 1 to a lactate curve (S1) displaying a genuine plateau during recovery from supramaximal exercise.

Table 2. Parameters of the fits of equation 1 to arterial lactate recovery curves from supramaximal exercises

Subject	La(O) $\mu mol \cdot l^{-1}$	A_1 $\mu mol \cdot l^{-1}$	A_2 $\mu mol \cdot l^{-1}$	γ_1 min^{-1}	γ_2 min^{-1}	La(∞) $\mu mol \cdot l^{-1}$	SD $\mu mol \cdot l^{-1}$	Absolute work rate		Relative work rate —— % $\dot{V}_{0_2}$ max
								W	$W \cdot kg^{-1}$	
S1	6,620	16,685	−23,080	0.227	0.0305	225	406	310	5.08	115
S2	4,740	13,767	−18,405	0.292	0.0263	102	242	305	4.83	109
S3	5,480	11,212	−16,270	0.306	0.0379	422	145	290	4.53	107
S4	5,730	11,154	−16,590	0.555	0.0371	294	86	320	4.78	107

SD = Standard deviation between the experimental and fitted data. La(∞) = computed lactate concentration at the end of recovery (t = infinity).

concentration were less than 1% of the maximal value (fig. 2). As for the trained subject (S4), after 3 min supramaximal exercise at 107% $\dot{V}_{0_2}$ max (4.78 $W \cdot kg^{-1}$), the arterial lactate recovery curve did not show such a plateau (fig. 3). All the arterial lactate recovery curves could be fitted by the biexponential time function of equation 1 (fig. 3, 4). The parameters of the fits are listed in table 2, where it is seen that the standard deviation between experimental and fitted curves ranged from 145 to 406 $\mu mol \cdot l^{-1}$ for untrained subjects, and was only 86 μmol $\cdot$ l^{-1} for the trained subject. It is also clear from table 2 that although S4 exercised at a higher absolute work rate, he displayed higher γ_1 than the untrained subjects. With respect to his γ_2, it was similar to that of one untrained subject (S3) but higher than those of the other two (S1 and S2).

Discussion

Lactate recovery curves from short muscular exercise usually show a biphasic evolution pattern composed of a rapidly increasing and a slowly decreasing phase. An important finding of the present study is that after 3 min supramaximal exercise of 107–115% $\dot{V}_{0_2}$ max (4.53–5.08 W $\cdot$ kg^{-1}), the arterial lactate recovery curves of untrained subjects displayed a pattern which appears to be specific to lactate recovery curves from exhaustive exercise. This particular evolution pattern (with a plateau between the well-established increasing and decreasing phases of the recovery curves) is also apparent in the data of other authors in similar experimental conditions [Greenhaff et al., 1988; Allsop et al., 1990; Itoh and Ohkuwa, 1990]. Therefore, it may correspond to a whole body status. As a consequence of the leveling off, the fits to the lactate recovery curves of the untrained subjects were slightly less good than to the classical biphasic evolution pattern. As already mentioned [Freund et al., 1986], the less-good fits of equation 1 to lactate recovery curves after supramaximal exercise do not constitute a shortcoming of the biexponential descriptive model, but result from pushing it to its limits of validity. It is nevertheless worth pointing out that, from a statistical point of view, the quality of the fits obtained for the untrained subjects was still reasonably good.

But the most interesting finding of the present study which is supplied by the use of the biexponential model is the low numerical values of γ_1 and γ_2 of the fits after 3 min supramaximal exercise (107–115% $\dot{V}_{0_2}$ max) in untrained subjects (S1, S2, and S3). According to the functional meaning given to γ_1 and γ_2 [Freund et al., 1986, 1989], namely the ability to exchange (γ_1) lactate between (M), the previously worked muscles, and (S), the remainder of the lactate space, and the overall ability (γ_2) to remove lactate during recovery, these results suggest that these abilities were severely impaired after supramaximal exercises.

However, does the dynamic information on lactate exchange and removal obtained from the fits to arterial lactate recovery curves have a physiological sense? Answering this question involves taking into account the following points: the reasons for the choice of the biexponential model; the validity of the two-compartment model as assessed from its predictive capacity; the comparison between the biexponential mathematical descriptive model to other descriptive models of lactate movements during the recovery; the justification of the functional meaning given to γ_1 and γ_2; and, finally, the possible relationships between lactate kinetics and some of the physico-biochemical modifications induced at the cellular level by exhaustive exercise.

Validity of the Biexponential Descriptive Model

The choice of a mathematical model in biology should not only satisfy the important criterion of goodness of fit. It should also repose on as simple a theory as possible and preferably one with which biologists are familiar. The biexponential model which has been used for the analytical description of lactate recovery curves fully satisfies these two criteria. On the one hand it supplies an accurate and concise description of lactate curves for the whole recovery period (coefficient of estimation $\geq$ 99% for recovery after submaximal to maximal exercise). It is worth mentioning that from a statistical point of view, any significant improvement in the goodness of the fits can be expected with additional exponential terms (at least those to lactate recovery curves after submaximal to maximal exercise). Therefore, this model appears, as pointed out by Carson et al. [1981], as a 'reduced model', i.e. one with the minimum of parameters. On the other hand, the theory of compartmental distribution [Cherruault and Loridan, 1977] can be employed in view of this biexponential behavior of the lactate concentration during recovery.

Concerning more specifically the utilization of the biexponential model even for the particular lactate pattern obtained after 3 min supramaximal exercises in untrained subjects, it should be noticed that the fits are nevertheless good, as already pointed out above. Furthermore, as mentioned in the 'Methods' section, it was not felt necessary to set an arbitrary distinction between submaximal and supramaximal exercises.

Predictive Capacity of the Biexponential Model

This aspect of the validity of the biexponential mathematical and two-compartmental distribution models has already been covered in several previous papers [Freund and Zouloumian, 1981a, b; Zouloumian and Freund, 1981a, b; Freund et al., 1984].

Comparison between the Biexponential Model and Other Descriptive Models of Lactate Movements during Recovery

Before entering into the actual comparison, it should be pointed out that the value of a model or a theory depends on the number of events or phenomena explainable by it. Good theories must aspire to have a very large explanatory capability [Hawking, 1989].

Recently, Peters-Futre et al. [1987] have studied lactate kinetics after five exercise bouts of 1-min duration each at 120% $\dot{V}_{0_2}$ max, separated by a period of 4 min rest. According to these authors, during the following partially *active* recovery, lactate curves were composed of three phases lasting, respectively, from 1–3 to 30 min for the first phase, from 30 to 45 min and from 45 to 90 min for the second and third phases. But during the following *passive* recovery, lactate curves were composed only of two phases lasting, respectively, from 1–3 to 45 min and from 45 to 90 min. Such decompositions of lactate recovery curves are arbitrary because they depend mainly on the time intervals of the sampling protocol. Moreover, the underlying reasons for this distinction between active recovery (3 phases) and passive recovery (2 phases) remain unclear. Although their data show, as already mentioned by several others [Gisolfi et al., 1966; Davies et al., 1970; Hermansen and Stensvold, 1972; Stamford et al., 1978; Bonen et al., 1979; Dodd et al., 1984], that during active recovery the removal of lactate is faster than during a passive one, this does not imply that the explanatory models are necessarily different. Rather, Freund and Zouloumian [1981a] have reported that on the same group of subjects the biexponential model (equation 1) can be used for describing the lactate evolution kinetics during active and passive recovery, and that the ability to remove lactate (γ_2) corresponding to active recovery was higher than that obtained during passive recovery. The good fits of equation 1 (fig. 5) to the data of Peters-Futre et al. [1987] are in agreement with the observations by Freund and Zouloumian [1981a], since they revealed that the γ_2 corresponding to active recovery (0.055 min^{-1}) was larger than that of passive recovery (0.046 min^{-1}).

Another point of importance is the consequences of the decomposition of lactate recovery curves made by Peters-Futre et al. [1987] on the conclusions drawn from their data. According to these authors, during the first phases of active and passive recovery, lactate removal rates, as assessed from the slopes of lactate curves, are higher than during the succeeding phases. But these higher rates should not be interpreted as a greater ability of the body to remove lactate during the first phases of active and passive recovery than during the succeeding ones. Such an interpretation would not fit the concept

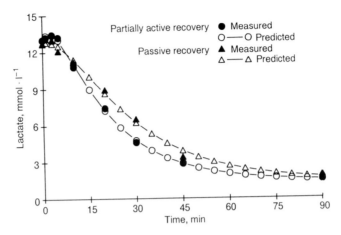

Fig. 5. Fits of equation 1 to Peters-Futre's lactate recovery data. During partially active and passive recovery the ability to remove lactate (γ_2) is, respectively, 0.055 and 0.046 min^{-1}.

of recovery after exhaustive exercise, since one might expect just the opposite, namely that the body would be more and more able to remove lactate as the recovery proceeds.

The explanation of the higher lactate removal rates during the first phases of active and passive recovery very likely lies in the large lactate concentrations prevailing at the end of exhaustive exercise. For example, assuming a lactate distribution volume of 35 liters for a subject of 70 kg, a blood lactate concentration of 15 mmol$\cdot$l^{-1} and a fractional turnover rate ranging within 0.02–0.03 min^{-1} at the onset of recovery, the calculated lactate removal rate [volume of lactate distribution (V, liters) times blood lactate concentration (La, mmol$\cdot$l^{-1}) times fractional lactate turnover rate (k, min^{-1}) [i.e. V$\cdot$La$\cdot$k] in this case would be 10.5–15.8 mmol$\cdot$min^{-1}. If later in the recovery the blood lactate concentration reaches 4 mmol$\cdot$l^{-1}, even with a higher fractional turnover rate of 0.05 min^{-1}, the lactate removal rate would only be 7 mmol$\cdot$min^{-1}, giving evidence of the importance of lactate concentration when looking at the lactate removal rate during recovery.

According to Carraro et al. [1989] the decline in blood lactate (from min 5 to min 60) after incremental exercise until exhaustion could fit a monoexponential model. The same analytical description had been used by Margaria et al. as early as 1933. But it should be observed that the very beginning of recovery is the end of exercise, and, at least for lactate studies, recovery goes on until blood lactate concentration has reached a steady basal level nearing that prevailing before the exercise. Thus, the reason why blood lactate concentrations were only considered between min 5 and min 60 of recovery remains unclear. Such a choice may have been a matter of convenience. But the rapidly increasing phase (t = 0 to t = 5 min), as well as the extreme latter part of lactate recovery curve with its more or less slow variation are not taken into account. It therefore appears that the biexponential model proposed by Freund and Gendry [1978] which accurately describes all lactate data from the end of exercise (t = 0) to the end of recovery (t = infinity), is much more general than the monoexponential model used by Margaria et al. [1933] and by Carraro et al. [1989].

Justification of the Functional Meaning Given to the Velocity Constants γ_1 and γ_2 of Equation 1

This question has already been approached in a previous paper of Freund et al. [1986]. Unfortunately there is no parameter to which our γ_1 can be compared, because as mentioned above, in the usually employed assessment of lactate kinetics data, the lactate space is assumed to be a single compartment. As for γ_2, briefly, it is formally similar to the lactate fractional turnover rate of the recovery [Freund et al., 1986]. Relative to the lactate resting fractional turnover which is known to range between 0.04 and 0.08 min^{-1} [Kreisberg et al., 1970; Searle and Cavalieri, 1972; Woll and Record, 1979], the γ_2 reached after 3 min easy to moderate exercise is enhanced (fig. 6). After strenuous exercise of 80–100 $\dot{V}_{O_2}$ max γ_2 reached approximately the lowest values reported for the lactate resting fractional turnover, while after 3 min supramaximal exercise at 107–115 $\dot{V}_{O_2}$ max, the values were decidedly smaller (fig. 6). As changes in γ_1 with work rate parallel those obtained for γ_2 (fig. 7), the comparison of the status of the body after submaximal to maximal exercise and supramaximal exercise may be represented as shown in figure 8. This figure illustrates the impaired lactate exchange and removal abilities after supramaximal exercise relative to rest and also relative to recovery after submaximal to maximal exercise.

In more recent works, Freund et al. [1989] and Fukuba [1990] reported that the relationships obtained by Stanley et al. [1985] and Mazzeo et al. [1986] between the lactate metabolic clearance rate during exercise and work rate displayed patterns similar to those between the lactate metabolic

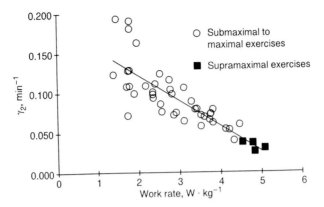

Fig. 6. Relationship between absolute work rate and the ability to remove lactate (γ_2). The higher the work rate, the lower the ability to remove lactate. Note that after supramaximal exercise, γ_2 is severely impaired; the values lie below those (0.04–0.08 min^{-1}) reported for the lactate resting fractional turnover. Data used for this figure are from Freund et al. [1986, 1989].

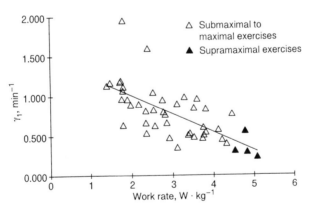

Fig. 7. Relationship between absolute work rate and the ability to exchange lactate (γ_1). The pattern is similar to that of the relationship between absolute work rate and γ_2. The γ_1 of the trained subject (solid symbol above the regression line) is higher than those of the untrained subjects (three solid symbols below the line). Data used for this figure are from Freund et al. [1986, 1989].

clearance rate of the recovery (computed by means of γ_2) and work rate of the preceding exercise. This similarity argues in favor of the functional meaning given to γ_2 and also indicates that some of the modifications induced by exercise are still perceptible during the following recovery. As pointed out by Freund et al. [1986, 1989], Oyono-Enguelle et al. [1989], recovery after exercise lower than 30% $\dot{V}_{O_2}$ max cannot be studied by the technique of the natural lactate load. For such exercises inducing very small or even no variation in blood lactate concentration, it would be difficult to fit the biexponential function of equation 1 to a lactate recovery curve that is hardly perceptible over the basal level. However, this method is well suited for maximal and even supramaximal exercises where injection or infusion techniques cannot be employed as was mentioned in the introductory section.

Another indirect justification of the functional meaning given to γ_1 and γ_2 appears through the comparison of the data of the trained subject and untrained ones. It is well known that training modifies the capillary blood supply to muscles [Bonen et al., 1978; Tesch et al., 1981]. Simon et al. [1986] mentioned that the time of appearance of the maximum lactate concentration occurred sooner for trained subjects than for untrained ones. Donovan and Brooks [1983] and Donovan and Pagliassotti [1990] reported further a higher lactate clearance in rats after endurance training. Recently, Oyono-Enguelle et al. [1990] demonstrated a higher ability to remove lactate in well-fit or well-trained athletes than in less-fit or less-trained ones. The results for the trained subject (S4) of the present study are in line with these previous observations because his higher γ_1 value (0.555 min^{-1} as compared to 0.227–0.306 min^{-1} for S1–S3), table 2) may be the result of a more developed capillary network in trained muscles. The higher γ_1 for the trained subject is in agreement with the data of Simon et al. [1986]. Also, the γ_2 value of 0.0371 min^{-1} obtained for S4 is of the same magnitude as the 0.0379 min^{-1} reached by S3, and higher than the γ_2 of the other two subjects (0.0263 and 0.0305 min^{-1}, respectively for S1 and S2). This result, including comparison between S3 and S4, agrees with those of Donovan and Brooks [1983] and Donovan and Pagliassotti [1990]. Indeed, based on the inverse dependence of γ_2 on the absolute work rate of the previously performed exercise [Freund et al., 1986, 1989], if S4 had exercised at the same absolute work rate of 290 W as S3 instead of at 320 W, he would have maintained a larger γ_2 during the following recovery than the 0.0371 min^{-1} obtained after exercise at 320 W. This line of reasoning is supported by the recent work by Oyono-Enguelle et al. [1990] who showed that the same subjects had higher γ_2 when they

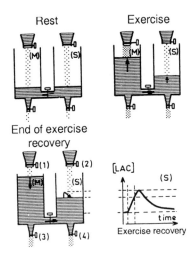

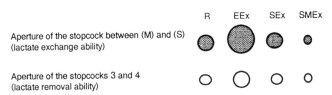

Fig. 8. Illustration of the body status in several conditions: R, EEx, SEx, SMEx, respectively, rest, recovery after easy exercise, strenuous submaximal to maximal exercise, and supramaximal exercise. The opening of the stopcocks between (M) and (S), from (M) and from (S) are proportional to the size of the symbol at the bottom right and are identified in this illustration to the lactate exchange (γ_1) and removal (γ_2) ability. 1 and 2 represent the stopcocks for lactate inflow (lactate production) in (M) and (S). This figure is a modification of figure 4 of Freund et al. [1986].

stopped cycling at 300 W than at 350 W. In the same way, it is very likely that a 3-min exercise at an absolute work rate higher than the 320 W of the present study would have more severely impaired the abilities of S4 to exchange and remove lactate with, as a consequence, the occurrence of a plateau of the arterial blood lactate comparable to those observed here for the other 3 subjects.

Possible Relationships between Lactate Kinetics and Some of the Physicochemical Modifications Induced at the Cellular Level by Exhaustive Exercise

Many studies have been conducted on exhaustive exercise and fatigue [Mutch and Banister, 1983; Edwards, 1983; Gibson and Edwards, 1985; Sjögaard, 1986; Sahlin, 1983, 1986; Wilkie, 1986; Völlestad and Sejersted, 1988; Kirkendall, 1990]. The review of some factors of peripheral fatigue [Kirkendall, 1990] strengthens our interpretation (supported by the low values of γ_1 and γ_2) that the abilities to exchange and remove lactate were impaired after supramaximal exercise.

Influence of Hydrogen Ions. It has been hypothesized [Kirkendall, 1990] that the known hydrogen ion accumulation during exhaustive exercise and its action at the cellular level may possibly be related to peripheral fatigue in the following numerous ways: decrease of membrane excitability [Ochardson, 1978], diminution of Ca^{2+} release from the sarcoplasmic reticulum [MacLaren et al., 1989]; increase of Ca^{2+} requirements for similar tension [Fabiato and Fabiato, 1978; Robertson and Kerrick, 1976]; inhibition of phosphorylase$_b$ kinase [Sahlin et al., 1975], of PFK [Danforth, 1965], of LDH and of several enzymes of the Krebs cycle [Bertocci and Gollnick, 1985]. In this list, the inhibitory effect of acidosis on LDH and on the Krebs cycle enzymes may be some of the underlying factors responsible for the impairment of lactate removal ability observed after supramaximal exercise. According to Brooks [1985] and to Peters-Futre et al. [1987], the main fate of lactate after exercise is its oxidative removal in the Krebs cycle. It then follows that during severe acidosis, as often recorded during exhaustive exercise, the first step of this process (transformation of lactate into pyruvate) and the other steps occurring later in the Krebs cycle ought to be impeded.

Acidosis has also been reported to enhance free radical formation [Jore and Perradini, 1988]. The possible consequence at the membrane level would be modification in its permeability, which could be reflected by impairment in lactate exchange ability.

Influence of Ammonia. Another factor of peripheral fatigue to which attention had to be paid when looking for underlying mechanisms of impaired lactate exchange and/or removal abilities during and after exhaustive exercise is ammonia. During strenuous exercise, production of ammonia is increased [for refs. see Weicker and Banister, 1990]. Owing to the low pH occurring during such situations, ammonia is converted into NH_4^+ [Mutch and Banister, 1983]. Ammonia and NH_4^+ accumulation at the cellular level has numerous repercussions on carbohydrate metabolism, namely: stimulation of PFK [Lowenstein, 1972], inhibition of PDH [Wircel and Erechinska, 1964] with, as a consequence, a locking effect on the pyruvate entering the Krebs cycle, inhibition of isocitrate dehydrogenase, one of the limiting enzymes of the Krebs cycle [Worcel and Erechinska, 1964] and finally inhibition of pyruvate carboxylase, one of the key enzymes of neoglucogenesis [MacLaren et al., 1989]. The last three of these actions might also account for the impaired lactate removal ability observed after supramaximal exercise.

Conclusion

From a practical point of view, the descriptive biexponential model appears to have more general applicability than other models used for analyzing lactate kinetics during recovery. Its capability of explaining tendencies occurring during recovery is high because the information obtained from it is coherent and plausible physiologically. This model constitutes per se a very useful heuristic tool. Although a reduced model (a minimum of parameters) [Carson et al., 1981], it has nevertheless permitted a structure and its mode of operating to be deduced. According to Freund and Gendry [1978], Freund and Zouloumian [1981a, b], Zouloumian and Freund [1981a, b], this structure may be assimilated with a two-compartment model of lactate distribution in the body. The biexponential mathematical model and the two compartment distribution have been shown to add to the information available on lactate recovery kinetics data. This model allows the estimation of internal functional parameters of the proposed structure. These internal parameters are of great interest in the physiological field, especially concerning lactate exchange processes which are hardly approachable using conventional tracer techniques.

Interested in the visibility and awareness of organism physiology, Evans et al. [1987] recently pointed out that '...each respective level (of research)* is being buried with information which makes it more difficult to visualize and optimize adjacent levels...' (*our parentheses). The results of the present study and their interpretation may be considered, on a practical basis, as an attempt to establish a cross-communication between research on the body (whole organism approach) and that at the molecular and cellular levels since as shown in the present paper, information delivered by arterial blood lactate recovery kinetics (impaired lactate exchange and removal abilities after supramaximal exercise) ought to be related to some of the predictable modifications at the cellular level during such situations.

References

Astrand, P.O.; Hultman, E.; Juhlin Dannfelt, A.; Reynolds, G.: Disposal of lactate during and after strenuous exercise in humans. J. appl. Physiol. *61:* 338–343 (1986).

Allsop, P.; Cheetham, M.; Brooks, S.; Hall, G.M.; William, C.: Continuous intramuscular pH measurement during the recovery from brief maximal exercise in man. Eur. J. appl. Physiol. *59:* 465–470 (1990).

Bertocci, L.A.; Gollnick, P.D.: pH effect on mitochondria and individual enzyme function. Med. Sci. Sports Exerc. *17:* suppl., p. 244 (1985).

Bonen, A.; Campbell, C.J.; Kirby, R.L.; Belcastro, A.N.: A multiple regression model for blood lactate removal in man. Pflügers Arch. *380:* 205–210 (1979).

Bonen, A.; Campbell, C.J.; Kirby, R.L.; Belcastro, A.N.: Relationship between slow-twitch muscle fibers and lactate acid removal. Can. J. Sport. Sci. *3:* 160–162 (1978).

Brooks, G.A.: The lactate shuttle during exercise and recovery. Med. Sci. Sports Exerc. *18:* 360–368 (1986).

Carraro, F.; Klein, S.; Rosenblatt, J.I.; Wolfe, R.R.: Effect of dichloroacetate on lactate concentration in exercising humans. J. appl. Physiol. *66:* 591–597 (1989).

Carson, E.R.; Cobelli, C.; Finkelstein, L.: Modeling an identification of metabolic systems. Am. J. Physiol. *240:* R120–R129 (1981).

Cherruault, Y.; Loridan, P.: Modélisation et méthodes mathématiques en biomédecine (Masson, Paris, 1977).

Cohen, R.D.: The production and removal of lactate; in Bossart, Perret (eds): Lactate in Acute Conditions. Int. Symp. Basel 1978, pp. 10–19 (Karger, Basel 1979).

Connor, H.; Woods, H.: Quantitative aspects of L(+) lactate in human beings. Metabolic acidosis, pp. 214–234 (Pitman Book Ltd., London/Ciba Foundation Symposium 87) (1982).

Danforth, W.H.: Activation of glycolytic pathway; in Chance, Estabrook, Control of Energy Metabolism, pp. 287–298. New York, Academic Press, (1965).

Davies, .T.M.; Knobbs, A.V.; Musgrove, J.: The rate of lactic acid removal in relation to different baselines of recovery exercise. Int. Z. Angew. Physiol. *28:* 155–161 (1970).

De Coster, A.; Denolin, H.; Messin, R.; Degre, S.; Vandermotten, P.: Role of the metabolites in the acid base balance during exercise; in Poortmans, Biochemistry of Exercise, pp. 5–34 (Karger, Basel 1969).

Diamant, B.; Karlsson, J.; Saltin, B.: Muscle tissue lactate after maximal exercise in man. Acta physiol. scand. *72:* 383–384 (1968).

Di Prampero, P.E.: Energetics of muscular exercise. Rev. Physiol. Biochem. Pharmacol. *89:* 143–222 (1981).

Dodd, S.; Powers, S.K.; Callender, T.; Brooks, E.: Blood lactate disappearance at various intensities of recovery exercise. J. appl. Physiol. *57:* 1462–1465 (1985).

Donovan, G.M.; Brooks, G.A.: Endurance training affects lactate clearance not lactate production. Am. J. Physiol. *244:* E83–E92 (1983).

Donovan, G.M.; Pagliassotti, M.J.: Enhanced efficiency of lactate removal after endurance training. J. appl. Physiol. 68: 1053–1058 (1990).

Edwards, R.H.T.: Biochemical basis of fatigue; in Knuttgen, Vogel, Biochemistry of Exercise, pp. 3–28. (Human Kinetics, Champaign 1983).

Evans, R.: Encouraging interaction among the biological branches (Letter to the Editor). Bone Mineral 2: 243–244 (1987).

Fabiato, A.; Fabiato. F.: Effects of pH on myofilaments and the sarcoplasmic reticulum of skinned cells from cardiac and skeletal muscles. J. Physiol. Lond. 276: 233–257 (1978).

Freminet, A.; Minaire, Y.: On the use of isotopic tracers for the study of lactate metabolism in vivo; in Jokl, Habbelinck (series eds.), Marconnet, Poortmans, Hermansen (vol. eds.), Physiological chemistry of training and detraining. Med. Sport Sci., vol. 17, pp. 25–39 (Karger, Basel 1984).

Freund, H.: Dosage automatique continu de l'acide pyruvique. Description de la méthode et application aux dosages simultanés de la pyruvicémie, de la lactatémie et de la glycémie. Ann. biol. Clin., Paris 25: 421–437 (1967).

Freund, H.; Gendry, P.: Lactate kinetics after short strenuous exercise in man. Eur. J. appl. Physiol. 39: 123–135 (1978).

Freund, H.; Jacquot, P.; Marbach, J.; Pellier, A.; Ramboarina, D.; Vogt, J.J.: In vivo experiments on the autoanalyser in a computerized environment; in Advances in automated analysis, vol. 1, pp. 195–198. Mediad Incorporated, White Plains (1970).

Freund, H.; Oyono-Enguelle, S.; Heitz, A.; Marbach, J.; Ott, C.; Zouloumian, P.; Lampert, E.: Work rate dependent lactate kinetics after exercise in humans. J. appl. Physiol. 61: 932–939 (1986).

Freund, H.; Oyono-Enguelle, S.; Heitz, A.; Marbach, J.; Ott, C.; Gartner, M.: Effect of exercise duration on lactate kinetics after short muscular exercise. Eur. J. appl. Physiol. 58: 534–542 (1989).

Freund, H.; Oyono-Enguelle, S.; Heitz, A.; Ott, C.; Marbach, J.; Gartner, A.; Pape, A.: Comparative lactate kinetics after short and prolonged submaximal exercise. Int. J. Sport Med. 11: 284–288 (1990).

Freund, H.; Zouloumian, P.: Lactate after exercise in man. I. Evolution kinetics in arterial blood. Eur. J. appl. Physiol. 46: 121–133 (1981a).

Freund, H.; Zouloumian, P.: Lactate after exercise in man. IV. Physiological observations and model predictions. Eur. J. appl. Physiol. 46: 161–176 (1981b).

Freund, H.; Zouloumian, P.; Oyono-Enguelle, S.; Lampert, E.: Lactate kinetics after maximal exercise in man; in Jokl, Hebbelinck (series eds.), Marconnet, Poortmans, Hermansen (vol. eds.), Physiological Chemistry of Training and Detraining. Med. Sport Sci., vol. 17, pp. 9–24 (Karger, Basel 1984).

Fukuba, Y.: Application of modeling to lactate kinetics in exercise. Ann. Physiol. Anthrop. 9: 203–211 (1990).

Gibson, H.; Edwards, R.H.T.: Muscular exercise and fatigue. Sports Med. 2: 120–132 (1985).

Gisolfi, C.; Robinson, S.; Turell, E.S.: Effects of aerobic work performed during recovery from exhausting work. J. appl. Physiol. 21: 1767–1772 (1966).

Greenhaff, P.L.; Gleeson, M.; Maugham, R.J.: The effects of diet on muscle pH and metabolism during high intensity exercise. Eur. J. appl. Physiol. 57: 531–539 (1988).

Harris, P.; Bailey, T.; Bateman, M.; Fitzgerald, M.G.; Gloster, J.; Harris, E.A.; Donald. K.W.: Lactate, pyruvate, glucose, and free fatty acid in mixed venous and arterial blood. J. appl. Physiol. 18: 933–936 (1965).

Harris, R.C.; Sahlin, K.; Hultman, E.: Phosphagen and lactate contents of muscle quadriceps femoris of man after exercise. J. appl. Physiol. *43:* 852–857 (1977).

Hawking, S.: Une brève histoire du temps. Du Big Bang au trou noir, pp. 28–29 (Flammarion, Paris, 1989).

Hermansen, L.; Stensvold, I.: Production and removal of lactate during exercise in man. Acta physiol. scand. *86:* 191–201 (1972).

Hermansen, L.; Maelhum, B.; Pruett, E.D.R.; Vaage, O.; Waldum, H.; Wessel-Aas, T.: Lactate removal at rest and during exercise; in Howald, Poortmans, Metabolic adaptation to physical exercise, pp. 101–105 (Birkhäuser, Basel 1975).

Hermansen, L.; Vaage, O.: Lactate disappearance and glycogen synthesis in human muscle after maximal exercise. Am. J. Physiol. *233:* E422–E429 (1977).

Heteneyi, G., Jr.; Perez, G.; Vranic, M.: Turnover and precursor-product relationships of nonlipid metabolites. Physiol. Rev. *63:* 606–667 (1983).

Holmgren, A.: Arterio-venous lactic acid differences in man at rest and during muscular work. Scand. J. clin. Lab. Invest. *11:* 150–153 (1959).

Itoh, H.; Ohkuwa, T.: Peak blood ammonia and lactate after submaximal and supramaximal exercise in sprinters and long distance runners. Eur. J. appl. Physiol. *60:* 271–276 (1990).

Jore, D.; Ferradini, C.: Peroxydation lipidique: rôle des radicaux libres et régulation par les vitamines E et C; in Douste-Blazy, Mendy, Biologie des lipides chez l'homme: De la physiologie à la pathologie, pp. 12–136 (Editions Médicales Internationales, Paris, 1988).

Karlsson, J.; Saltin, B.: Lactate, ATP, and CP in working muscle during exhaustive exercise. J. appl. Physiol. *29:* 589–602 (1970).

Karlsoson, J.: Lactate and phosphagen in working muscle of man. Acta physiol. scand. *358:* suppl., pp 1–72 (1971).

Karlsson, J.; Diamant, B.; Saltin, B.: Muscle metabolites during submaximal and maximal exercise in man. Scan. J. clin. Lab. Invest. *26:* 385–394 (1971).

Kirkendall, D.T.: Mechanism of peripheral fatigue. Med. Sci. Sports Exerc. *22:* 444–449 (1990).

Kreisberg, R.A.; Pennington, L.F.; Bashell, C.P.: Lactate turnover and gluconeogenesis in normal obese humans. Effect of starvation. Diabetes *19:* 53–63 (1970).

Layzer, R.B.; Rowland, L.P.; Bank, W.J.: Physical and kinetics properties of human phosphofructokinase from skeletal muscle and erythrocytes. J. biol. Chem. *244:* 3823–3831 (1969).

Lowenstein, J.M.: Ammonia production in muscles and other tissues: The purine nucleotide cycle. Physiol. Rev. *52:* 382–414 (1972).

MacLaren, D.P.M.; Gibson, N.; Parry Billings, M.; Edwards, R.H.T.: A review of metabolic factors in fatigue. Exerc. Sport Sci. Rev. *17:* 29–68 (1989).

Margaria, R.; Edwards, H.T.: The removal of lactic acid from the body during recovery from muscular exercise. Am. J. Physiol. *107:*681–686 (1934).

Margaria, R.; Edwards, H.T.; Dill, D.B.: The possible mechanism of contracting and paying the oxygen debt and the role of lactic acid in muscle contraction. Am. J. Physiol. *106:* 689–715 (1933).

Mazzeo, R.S.; Brooks, G.A.; Schoeller, D.A.; Budinger, T.F.: Disposal of 1-13C lactate in humans during rest and exercise. J. appl. Physiol. *60:* 232–241 (1986).

Minaire, Y.; Studievic, C.; Foucherand, F.: Dosage automatique de l'acide lactique. Possibilités d'utilisation en continu. Path. Biol. *13* 1170–1173 (1965).

Mitchell, A.M.; Cournand, A.: The fate of circulating lactic acid in the human lung. J. clin. Invest. *34:* 471–476 (1955).

Mutch, B.A.; Banister, E.W.: Ammonia metabolism in exercise and fatigue: a review. Med. Sci. Sports Exerc. *15:* 41–50 (1983).

Norwich, K.H.; Wassermann, D.H.: Calculation of metabolic fluxes with anatomically separated sources of tracer and tracee. Am. J. Physiol. *25:* 715–720 (1986).

Ochardson, P.: The generation of nerve impulse in mammalian axons by changing concentrations of the normal constituents of extracellular fluid. J. Physiol. *275:* 177–189 (1978).

Okajima, F.; Chenoweth, M.; Roghstad, R.; Dunn, A.; Katz, J.: Metabolism of ^3H and ^{14}C labeled lactate in starved rats. Biochem. J. *194:* 525–540 (1981).

Olsen, C.; Strange Petersen, E.: The lactate/pyruvate ratio in muscular work and following injection of lactate in man. Pflügers Arch. *342:* 359–365 (1973).

Oyono-Enguelle, S.; Gartner, M.; Marbach, J.; Heitz, A.; Ott, C.; Freund, H.: Comparison of arterial and venous blood lactate kinetics after short exercise. Int. J. Sports Med. *10:* 16–24 (1989).

Oyono-Enguelle, S.; Marbach, J.; Heitz, A.; Ott, C.; Gartner, M.; Pape, A.; Vollmer, J.C.; Freund, H.: Lactate removal ability and graded exercise in humans. J. appl. Physiol. *68:* 905–911 (1990).

Peters-Futre, E.M.; Noakes, T.D.; Raine, R.I.; Terblanche, S.E.: Muscle glycogen repletion during active post exercise recovery. Am. J. Physiol. *253:* E305–E311 (1987).

Poortmans, J.R.; Delescaille-Vanden Bossche, J.; Leclercq, R.: Lactate uptake by inactive forearm during progressive leg exercise. J. appl. Physiol. *45:* 835–839 (1978).

Robertson, S.; Kerrick, W.: The effect of pH on submaximal calcium ion activated tension in skinned frog skeletal fibers. Biophys. J. *16:* 17A (1976).

Sahlin, K.: Effect of acidosis on energy metabolism and force generation in skeletal muscle; in Knuttgen, Vogel, Biochemistry of exercise, pp. 151–160 (Human Kinetics, Champaign 1983).

Sahlin, K.: Muscle fatigue and lactic acid accumulation. Acta physiol. scand. *128:* suppl. 556, pp. 83–91 (1986).

Sahlin, K.; Harris, R.C.; Hultman, E.: Creatine kinase equilibrium and lactate content compared with muscle pH in tissue samples obtained after isometric contraction. Biochem. J. *152:* 173–180 (1975).

Sahlin, K.; Harris, R.C.; Nylind, B.; Hultman, E.: Lactate content and pH in muscle samples obtained after dynamic exercise. Pflügers Arch. *367:*143–149 (1976).

Searle, G.L.; Cavalieri, R.R.: Determination of lactate kinetics in the analysis of data from single injection versus continuous infusion methods. Proc. Soc. exp. Biol. Med. *139:* 1002–1006 (1972).

Sejersted, O.M.; Medbö, J.I.; Hermansen, L.: Metabolic acidosis and changes in water balance after maximal exercise; in Metabolic acidosis, Ciba Found. Symp. 87, pp. 157–167 (Pitman, London, 1982).

Sejersted, O.M.; Medbö, J.I.; Orheim, A.; Hermansen, L.: Relationship between acid base status and electrolyte balance after maximal work of short duration; in Jokl, Hebbelinck (series eds.), Marconnet, Poortmans, Hermansen (vol. eds.), Physiological chemistry of training and detraining. Med. Sport Sci., vol. 17, pp. 40–55 (Karger, Basel 1984).

Simon, J.; Young, J.L.; Blood, D.K.; Segal, K.R.; Case, R.B.; Gutin, B.: Plasma lactate and ventilatory thresholds in trained and untrained cyclists. J. appl. Physiol. *60:*777–781(1986).

Sjögaard, G.: Water and electrolyte shifts during exercise and their relation to muscular fatigue. Acta physiol. scand. *128:* suppl. 556, pp. 129–136 (1986).

Stamford, B.A.; Moffatt, R.J.; Weltman, H.; Maldonado, L.; Curtis, M.: Blood lactate disappearance after supramaximal one legged exercise. J. appl. Physiol. *45:* 244–248 (1978).

Stanley, W.C.; Gertz, E.W.; Wisneski, J.A.; Morris, D.L.; Neese, R.A.; Brooks, G.A.: Systemic lactate kinetics during graded exercise in man. Am. J. Physiol. *249:* 595–602 (1985).

Tesch, P.; Charp, D.S.; Daniels, W.L.: Influence of fiber type, fiber composition and capillary density on onset of blood lactate accumulation. Int. J. Sports Med. *2:* 252–255 (1981).

Völlestad, N.K.; Sejersted, O.M.: Biochemical correlates of fatigue. Eur. J. appl. Physiol. *57:* 336–347 (1988).

Weicker, H.; Banister, E.W.: Purine nucleotide cycle and ammonia formation. Int. J. Sport Med. *11:* suppl. 2, pp. S35–S142 (1990).

Wilkie, D.R.: Muscular fatigue effects of hydrogen ions and inorganic phosphate. Fed. Proc. *45:* 2921–2933 (1986).

Woll, P.J.; Record, C.O.: Lactate elimination in man: Effect of lactate concentration and hepatic dysfunction. Eur. J. clin. Invest. *9:* 397–404 (1979).

Worcel, A.; Erecinska, M.: Mechanism of inhibitory action of ammonia on the respiration of rat liver mitochondria. Biochem. biophys. Acta *65:* 27–33 (1964).

Zouloumian, P.; Freund, H.: Lactate after exercise in man. II. Mathematical model. Eur. J. appl. Physiol. *46:* 135–147 (1981a).

Zouloumian, P.; Freund, H.: Lactate after exercise in man. III. Properties of the compartment model. Eur. J. appl. Physiol. *46:* 149–160 (1981b).

S. Oyono-Enguelle, Laboratoire de Physiologie Appliquée, Faculté de Médecine, 4, rue Kirschleger, F–67085 Strasbourg Cedex (France)

Marconnet P, Komi PV, Saltin B, Sejersted OM (eds): Muscle Fatigue Mechanisms in
Exercise and Training. Med Sport Sci. Basel, Karger, 1992, vol 34, pp 162–171

Do Muscle Sensory Receptors Compensate for or Contribute to Neuromuscular Fatigue?[1]

Robert S. Hutton, Sally W. Atwater, Daniel L. Nelson

Department of Psychology, University of Washington, Seattle, Wash., USA

Surprisingly, little research has been conducted on the role of muscle sensory receptors during neuromuscular fatigue. Whereas myogenic contractile failure and possible causal mechanisms have been well documented, the typical experimental protocols used do not allow insight into events that may be occurring in neural circuitry serving alpha motor neurons (MNs) or 'the final common path'. In this review, spinal segmental and intersegmental reflexes and their adaptive responses that occur during skeletal muscle fatigue will be addressed.

The following observations summarize findings that have been reported over the past 20 years concerning muscle fatigue at the level of the motor unit (MU). These include: (1) excitation-contraction coupling failure attributed to myogenic metabolic mechanisms [5, 12, 29, 30]; (2) adaptive changes in electrotonic neuronal and sarcolemmal membrane properties [1, 3, 20, 21, 26, 35]; (3) pre-junctional failure in selected MU types [5, 10]; (4) a decline in discharge frequencies in alpha MNs [2, 26, 34]; (5) a shift in the force-frequency relationship toward lower frequencies [1, 12, 24, 26, 34]; (6) a decrease in the mean frequency power spectrum of the surface electromyographic activity (EMG) [1, 3, 23, 24]; and (7) an increase or decrease in the surface EMG amplitude as muscle force declines [1, 3, 36].

This paper will focus on the possible role(s) of muscle sensory receptors in contributing to the 2nd through the 7th observations listed above. For the sake of brevity, documentation of statements made will be in the form of review articles whenever possible.

[1] This paper was partially funded by the University of Washington, Graduate Research Science Fund-Biological Research Science Grant (NIH) and a grant from the American College of Sports Medicine Foundation.

The first attempt to associate the role of muscle sensory receptors during muscle fatigue, at the electrophysiological level, appears to have been conducted by Bronk [J Physiol (Lond) 1929;67:270–281]. He studied the changes in frequency of unidentified afferent nerve impulses from muscle during periods of prolonged loading or during successive stretches. During prolonged loading over several minutes, the discharge frequency was found to decrease, whereas, during successive stretches (over several thousand times with only 1-second intervals) no appreciable effects on discharge frequency were observed. The latter observation agrees with unpublished findings observed in our laboratory. The former finding is equivocal, as sensory receptors are known to adapt to a constant adequate stimulus with a decline in frequency, but a brief removal of the stimulus allows a return to control values in response to the same stimulus. This instantaneous recovery is not typically characteristic of a fatigue-induced response.

With the elucidation of the stretch reflex and indirect control of alpha motoneurons by way of the gamma-loop through alpha-gamma coactivation [38], considerable speculation followed as to whether the gamma-loop offered a pathway of gain control of alpha motoneurons during neuromuscular fatigue [9, 27, 31, 39]. Marsden et al. [27] addressed the issue, indirectly, by having human subjects perform flexion movements of the interphalangeal joint of the thumb against a constant torque load. Subjects followed visually displayed movement trajectories while electromyographic activity (EMG) and joint displacement were recorded. During the course of the movement, an unexpected increase or decrease in the resistive torque either stretched the thumb, halted its forward progression, or increased its rate of displacement. Increases and decreases in the resistive torques caused a short latency (about 50 ms) reflexive increase and decrease in the EMG amplitudes, respectively. The gain in EMG appeared to be related to the level of muscle activation as determined by the voluntary effort of the subjects. The investigators concluded that in fatigued muscle '... gain is boosted as the muscle has to be activated more strongly to keep up with the same force'.

However, these findings do not provide direct evidence that either group Ia or II muscle spindle receptors provide an increase in excitatory current to alpha motoneurons during muscle fatigue.

To our knowledge, the first studies of muscle spindle responses during muscle fatigue were conducted by Nelson and Hutton [31]. These studies were prompted by the earlier findings that brief periods of muscle contraction increased the resting, dynamic, and static frequency responses of, principally, Ia receptors [16, 19, 33]. However, it was unclear whether these

Table 1. Percent change in firing frequency of muscle spindle afferents during muscle fatigue

Adaptation	Receptor[1]	
	Ia	II
% increase (n)	189 (25)	167(18)
% decrease (n)	111(9)	44(4)
% change (n)	+78(49)*	+78(33)*

*p<0.01.
[1] Mean control frequencies (imp/s) were: Ia = 7; II = 9.

same adaptations would occur if the duration of muscle contraction were long enough to induce muscle fatigue.

Therefore, studies were conducted on the surgically isolated cat gastrocnemius muscle. Muscle spindle afferent fibers were isolated in the L7 and S1 dorsal roots and categorized on the basis of conventional criteria (e.g. conduction velocity, frequency response to stretch). Muscle fatigue was induced by stimulating the distal ends of the transected L7 and S1 ventral roots at a sustained frequency of 100 Hz, 1.7 times twitch threshold. Resting, dynamic, and static discharge and vibration frequencies (200 Hz) were recorded before and during muscle fatigue. A stretch of 5 and 25 mm/s over a 5-mm displacement was imposed to record the ramp stretch responsivity of the afferent receptors monitored. Response latency to stretch was also quantified. Since the findings were identical when comparing the spindle responses under both conditions of stretch, only the data for the slow stretch condition will be discussed.

Changes in resting discharge before and after fatigue are shown in table 1. The resting levels of firing frequency of both receptors were elevated by 78%. The percent change and number falling in the group II fiber category that increased their discharge were far greater than previous accounts observed in unfatigued muscle. The functional consequence of this change would be to elevate tonic levels of neural drive to ongoing muscle tension.

The changes in dynamic sensitivity were modest (13 and 15%, respectively) when considered as the mean frequency over the course of a 5-mm stretch (table 2). However, the main change in frequency occurred during the first 3 mm of stretch. The theoretical basis for this will be explained later. As

Table 2. Percent change in dynamic firing frequency of muscle spindle afferents to ramp stretch (5 mm/s) during muscle fatigue

Adaptation	Receptor[1]	
	Ia	II
% increase (n)	28(31)	31(22)
% decrease (n)	19(9)	18(7)
% change (n)	+13(49)*	+15(34)*

*p<0.01.
[1] Mean control prestimulus frequences (imp/s) were: Ia = 47; II = 39.

Table 3. Percent change in static firing frequency of muscle afferents to ramp stretch (5 mm/s) during muscle fatigue

Adaptation	Receptor[1]	
	Ia	II
% increase (n)	25(11)	21(13)
% decrease (n)	32(31)	16(14)
% change (n)	−16(49)*	0(34)

*p<0.01.
[1] Mean control prestimulus frequencies (imp/s) were: Ia = 44; II = 43.

a further indication of dynamic sensitivity, vibration frequencies were found to be significantly elevated (p<0.01) in both populations (+89 and +37 Hz in Ia and II receptors, respectively).

Interestingly, changes in static sensitivity were suppressed in group Ia fibers and showed no net change in the II afferent fibers (table 3; these results will also be explained later). A marked difference was seen in the response latency of both receptors. The recruitment or rise in resting tonic frequency levels at the time of stretch onset were advanced by −174 and −163 ms, respectively (table 4). In all these observations, it is important to note that they can be uncoupled by muscle stretch, suggesting a mechanical mechanism as the basis for the changes in spindle discharge.

Table 4. Percent change in response latency of muscle afferents to ramp stretch (5 mm/s) during muscle fatigue

Adaptation	Receptor[1]	
	Ia	II
% increase (n)	8(1)	72(2)
% decrease (n)	93(40)	73(30)
% change (n)	−73(49)*	−62(34)*

*p<0.01.
[1]Mean control latencies were: Ia = 237 ms; II = 263 ms.

The passive peak force-time responses of the whole muscle during fatigue were elevated at both stretch velocities (p<0.01, mean increase =1.1 N at both stretch velocities). The 'short range stiffness' response [32] accounted for a greater range of the total force-time relationship during fatigue. These alterations in mechanical properties of extrafusal muscle fibers had to be offset by similar changes in intrafusal muscle fibers, otherwise one would expect an unloading effect on the muscle spindle receptors and a net decrease in discharge frequency instead of an increase. As noted earlier, the direction of these frequency and latency responses can be induced simply through increased activation of proprioceptive pathways [16, 19, 33]. However, during fatigue all changes appeared amplified.

The causal mechanism most likely resides in a contraction-induced increase in intrafusal muscle stiffness through the persistence of actin and myosin binding [7, 8, 13, 19], as has been shown in extrafusal muscle [15]. Early adaptations may also involve a short period of afterdischarge in gamma motoneurons [25]. As further evidence for a myogenic mechanism, apart from the fact that the enhanced responses can be reset to precontraction control values by muscle stretch (which is thought to break residual actin and myosin bonds), these findings could not be replicated by inducing fatigue at a stimulus intensity below the threshold of gamma efferents. A myogenic explanation would also account for why the increased dynamic response occurred only through the first 3 mm of stretch and the static response was unaffected, i.e. a 5-mm stretch was sufficient to decouple postcontraction-induced actin-myosin binding.

These findings suggest that there is an initial increase in the gain of the stretch reflex. They are also in support of the increased gain seen in EMG during a stretch perturbation as reported by Marsden et al. [27]. Contraction-induced increases in muscle spindle stretch sensitivity have since been reported in humans [11]. Moreover, Christakos and Windhorst [9] and their colleagues found in cat gastrocnemius muscle that during the course of muscle fatigue the gain of the 'subsystem transforming force' to afferent discharge increased, suggesting that the overall gain remains relatively high between motor units and spindle afferents. Subsequent studies confirmed an increase in gain in the frequency domain [39].

An issue that remains open to debate is whether intrafusal muscle fibers can eventually be driven to fatigue in a like manner as extrafusal muscle fibers. Bongiovanni and Hagbarth [6] recently reported indirect evidence that intrafusal muscle may be fatigued during the later stages of the decline in extrafusal muscle force. They superimposed a high frequency vibration (a relatively selective stimulus to Ia receptors) on dorsiflexors of human subjects during a sustained MVC and found an initial facilitation in EMG followed by a decline in EMG in parallel with muscle force. The decline in EMG was presumed associated with intrafusal muscle fatigue. We find this suggestion unlikely due to the relatively short time to induce fatigue but the possibility remains open for further experimentation. For example, Hutton and Doolittle [18] have found tonic vibration reflexes (TVRs) to be depressed in highly trained endurance runners following a maximum oxygen capacity treadmill test, whereas TVRs were potentiated in sedentary controls.

With respect to muscle proprioceptors, the next question to be addressed was whether the Golgi tendon organ reflex, likewise, contributed to the compensation of neuromuscular fatigue through, for example, disinhibition. Using the same test protocol as for the muscle spindle data, Hutton and Nelson [17] observed changes in Ib afferents before and during muscle fatigue. The results of their findings are qualitatively summarized in table 5.

Table 5. Summary of Ib discharge in fatigued muscle (n = 41)

Control condition	Adaptation
Resting discharge	decreased (if firing)
Static response	decreased
Vibration response	decreased
Dynamic response (peak)	decreased
Dynamic sensitivity (x frequency)	decreased
Response latency (x)	increased

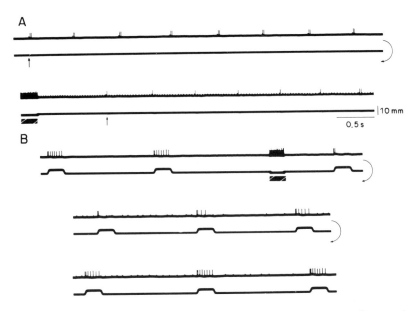

Fig. 1. A tendon organ Ib frequency response (top trace) to gastrocnemius muscle twitch before *(A)* and following *(B)* muscle fatigue. Displacement at low gain is shown on the bottom trace (downward deflection denotes muscle shortening) in response to a 5-mm stretch at a constant velocity of 250 mm/s or an equivalent stretch velocity to approximate the time to peak tension. Note the selective effect on the static response following muscle fatigue. All traces are continuous. See text for further explanation.

In all cases, except the static response to stretch, the frequency responses of Ib receptors to a 5-mm stretch at 5 mm/s were the inverse of those seen in muscle spindle receptors. The clear and consistent finding was that prolonged muscle contraction caused desensitization of Ib receptors to stretch and vibration. However, these observations were less marked when the test stimulus pulse elicited a muscle twitch or when the stretch velocity was adjusted to match that which would occur during the time to peak tension [36]. As shown in figure 1A, the prefatigue frequency response of this Ib receptor shows a doublet response to muscle twitch. During muscle fatigue, only a single spike potential is seen over a period of about 4 s. Moreover, when the 5-mm ramp stretch was adjusted to match the time to peak tension, the rather surprising observation was a selective depression of the static response with a sluggish recovery of the static response over a period of about 6 s. Perhaps this accounts for some of the earlier observations reported by Bronk as cited earlier. Nevertheless, these less marked adaptations during muscle fatigue would contribute to disinhibition to the muscles undergoing excitation to a motor command.

Recent experiments in our laboratory on human Ib afferent input during muscle fatigue, suggest that, indeed, the transynaptic transmission of Ib afferent input may be blunted during muscle fatigue [unpubl. observations]. Therefore, proprioceptive muscle receptors appear to serve a compensatory excitatory function during muscle fatigue.

However, recent evidence [14] has been reported to show that the cat soleus muscle force significantly decreases, when the fatigued medial gastrocnemius muscle is either stretched or the muscle nerve electrically stimulated. Based, in part, on the evidence presented above, these investigators proposed that the origin of the rise in reflexive inhibitory input was most likely caused by group III and IV afferent receptors. They also proposed that this inhibitory input may account for the decrease in motor unit firing rates as neuromuscular fatigue ensues [2, 26, 34].

How does one reconcile these diametric spinal reflex responses caused by group I and II receptors, on the one hand, and presumably group III and IV receptors on the other? Given the likely function of group III and IV pathways in mediating pain [28], we propose that increased feedback from small afferents may be serving a protective function, i.e. contributing to a balance between excitatory and inhibitory current which might eventually lead to withdrawal or flexion reflex due to pain. Recall that the inhibitory reflex was reported during fatigue in an extensor muscle. To our knowledge, a similar inhibitory reflex during fatigue has not been demonstrated in flexor muscle. In contrast, muscle proprioceptors compensate for fatigue by providing additional excitatory drive to the motor command.

In particular, one would expect this to be the case in highly motivated subjects [1, 5, 10]. A test of the validity of this proposal awaits further experimentation.

References

1   Bigland-Ritchie B: EMG/force relations and fatigue of human voluntary contributions. Exerc Sport Sci Rev 1981;9:75–117.
2   Bigland-Ritchie B, Dawson R, Johansson NJ, Lippold OCJ: Reflex origin for the slowing of motoneurone firing rates in fatigue of human voluntary contractions. J Physiol (Lond) 1986;378:451–459.
3   Bigland-Ritchie B, Johansson R, Lippold OCJ, Woods JJ: Contractile speed and EMG changes during fatigue of sustained maximal voluntary contractions. J Neurophysiol 1983;50:313–324.

4   Bigland-Ritchie B, Kukulka CG, Lippold OCJ, Woods JJ: The absence of neuromuscular transmission failure in sustained maximal voluntary contractions. J Physiol (Lond) 1982;330:265–278.

5   Bigland-Ritchie B, Woods JJ: Changes in muscle contractile properties and neural control during human muscular fatigue. Muscle Nerve 1984;7:691–699.

6   Bongiovanni LG, Hagbarth K-E: Tonic vibration reflexes elicited during fatigue from maximal voluntary contractions in man. J Physiol 1990;423:1–14.

7   Brown MC, Goodwin GM, Matthews PBC: After-effects of fusimotor stimulation on the response of muscle spindle primary afferent endings. J Physiol (Lond) 1969;205: 677–694.

8   Brown MC, Goodwin GM, Matthews PBC: The persistence of stable bonds between actin and myosin filaments of intrafusal muscle fibers following their activation. J Physiol (Lond) 1970;210:9–10P.

9   Christakos CN, Windhorst U: Spindle gain increase during muscle unit fatigue. Brain Res 1986;365:333–392.

10 Edgerton VR, Hutton RS: Nervous system and sensory adaptation; in Bouchard et al (eds): Exercise, Fitness, and Health. Champaign, Human Kinetics Press, 1988, pp 363–376.

11 Enoka RM, Hutton RS, Eldred E: Change in excitability of tendon tap and Hoffmann reflexes following voluntary contraction. EEG Clin Neurophysiol 1980;48: 664–672.

12 Faulkner JA: Fatigue of skeletal muscle fibers; in: Hypoxia, Exercise, and Altitude: Proc Third Banff Int Hypoxia Symp. New York, Liss, 1983, pp 243–255.

13 Gregory JE, Prochazka A, Proske U: Response of muscle spindles to stretch after a period of fusimotor activity in freely moving and anaesthetized cats. Neurosci Lett 1977;4:67–72.

14 Hayward L. Brietbach D, Rymer WZ: Increased inhibitory effects of close synergists during muscle fatigue in the decerebrate cat. Brain Res 1988;440:199–203.

15 Hill DK: Tension due to interaction between the sliding filaments in resting striated muscle. The effect of stimulation. J Physiol (Lond) 1968;199:637–684.

16 Hutton RS: Acute plasticity in spinal segmental pathways with use: Implications for training: in Kimamoto M (ed): Neural and Mechanical Control of Movement. Kyoto, Yamaguchi Shoten, 1984.

17 Hutton RS, Nelson DL: Stretch sensitivity of Golgi tendon organs in fatigued gastrocnemius muscle. Med Sci Sports Exerc 1986;18:69–74.

18 Hutton RS, Doolittle TL: Resting electromyographic triceps surae activity and tonic vibration reflexes in subjects with high and average-low maximum oxygen uptake capacities. Res Q Exerc Sports 1987;58:280–285.

19 Hutton RS, Smith JL, Eldred E: Post-contraction sensory discharge from muscle and its source. J Neurophysiol 1973;36:1090–1103.

20 Kernell D: Organization and properties of spinal motoneurons and motor units. Prog Brain Res 1986;64:21–30.

21 Kernell D, Monster AW: Motoneurone properties and motor fatigue. Exp Brain Res 1982;46:197–204.

22 Kirsch RF, Rymer WZ: Neural compensation for muscular fatigue: Evidence for significant force regulation in man. J Neurophysiol 1987;57:1893–1910.

23 Komi PV, Tesch : EMG frequency spectrum, muscle structure, and fatigue during dynamic contractions in man. Eur J Appl Physiol 1979;42:41–50.

24 Kranz H, Cassell JF, Inbar GF: Relation between electromyogram and force in fatigue. J Appl Physiol 1985;59:821–825.
25 Kuffler SW, Hunt CC, Quilliam JP: Function of medullated small-nerve fibers in mammalian ventral roots: Efferent muscle spindle innervation. J Neurophysiol 1951;14:29–54.
26 Marsden CD, Meadows JC, Merton PA: "Muscular wisdom" that minimizes fatigue during prolonged effort in man: Peak rates of motoneuron discharge and slowing of discharge during fatigue; in Desmedt JE (ed): Motor Mechanisms in Health and Disease. New York, Raven Press, 1983, pp 169–211.
27 Marsden CD, Merton PA, Morton HB: Servo action in the human thumb. J Physiol (Lond) 1976;257:1–44.
28 Mense S, Meyer H: Different types of slowly conducting afferent units in cat skeletal muscle and tendon. J Physiol (Lond) 1985;363:403–417.
29 Merton PA: Problems of muscular fatigue. Br Med Bull 1956;12:219–221.
30 Nassare-Gentina V, Passonneau JV, Vegara JL, Rapoport SE: Metabolic correlates of fatigue and of recovery from fatigue in single frog muscle fibers. J Gen Physiol 1978;72:593–606.
31 Nelson DL, Hutton RS: Dynamic and static stretch responses in muscle spindle receptors in fatigued muscle. Med Sci Sports Exerc 1985;17:445–450.
32 Rack PMH, Westbury DR: The short range stiffness of active mammalian muscle. J Physiol (Lond) 1972;229:16–17P.
33 Smith JL, Hutton RS, Eldred E: Post-contraction changes in sensitivity of muscle afferents to static and dynamic stretch. Brain Res 1974;78:193–202.
34 Smith S, Woods JJ: Changes in motoneuron firing rates during sustained maximal voluntary contractions. J Physiol (Lond) 1983:335–356.
35 Stephens JA, Taylor A: Fatigue of maintained voluntary contraction in man. J Physiol (Lond) 1972;220:1–18.
36 Stuart DG, Goslow GE, Mosher CG, Reinking RM: Stretch responsiveness of Golgi tendon organs. Exp Brain Res 1970;10:463–476.
37 Suzuki S, Hutton RS: Postcontractile motoneuronal discharge produced by muscle afferent activation. Exp Neurol 1983;81:141–152.
38 Vallbo AB: Muscle spindle response at the onset of isometric voluntary contraction in man: Time difference between fusimotor and skeletomotor effects. J Physiol (Lond) 1974;218:405–431.
39 Windhorst U, Christakos CN, Koehler W, Hamm TM, Enoka RM, Stuart DG: Amplitude reduction of motor unit twitches during repetitive activation is accompanied by relative increase of hyperpolarizing membrane potential trajectories in homonymous α-motoneurons. Brain Res 1986;398:181–184.

Prof. Robert S. Hutton, Department of Psychology, NI-25,
University of Washington, Seattle, WA 98195 (USA)

Marconnet P, Komi PV, Saltin B, Sejersted OM (eds): Muscle Fatigue Mechanisms in
Exercise and Training. Med Sport Sci. Basel, Karger, 1992, vol 34, pp 172–181

Neuromuscular Fatigue during Repeated Stretch-Shortening Cycle Exercises

Paavo V. Komi[a], *Caroline Nicol*[a], *Pierre Marconnet*[b]

[a]Department of Biology of Physical Activity, University of Jyväskylä, Finland;
[b]Laboratory of Biomechanics and Biology of Exercise,
University of Nice-Sophia Antipolis, Nice, France

As discussed in the various chapters of this volume, fatigue has been studied mostly under very strict conditions of either isometric or concentric actions. In natural locomotion, however, muscle function takes place primarily in stretch-shortening cycles (SSC) where eccentric and concentric parts of the muscular work follow each other [Komi, 1984]. Running, jumping and hopping are examples of activities in which SSC-type muscle function can be identified for leg extensor muscles. When direct Achilles tendon forces have been measured during this type of activity in human [Komi, 1990] or animal [Gregor et al., 1987] experiments the mechanical (and metabolic) performance of the skeletal muscle was modified in the concentric part of the SSC. For example, when the in vivo force-velocity (F-V) curve of the SSC is compared to that of the classical F-V curve of the isolated concentric action, SSC situation demonstrates a clear performance potentiation. This potentiation after active prestretch (eccentric action) is a complex phenomenon and can be explained by a variety of elastic and chemomechanical events [Huijing, 1991]. In these discussions, the importance of the reflex influences must also be taken into consideration. When an active muscle is put under stretch, the activity from both the muscle spindles and the Golgi tendon organs will determine which of these reflexes – facilitatory or inhibitory – will dominate, and what magnitude of the performance potentiation from the reflex sources could be. Hoffer and Andreassen [1981] have documented well that when the reflexes are intact, the muscle stiffness is greater per the same operating force than in an areflexive muscle. Thus, the reflexes may make a net contribution to muscle stiffness during the eccentric part of SSC.

SSC is thus a form of muscle function where all the major components of performance 'sources' are loaded: neural, mechanical (elastic), and metabolic. The purpose of the present report is to characterize how SSC-type exercises are susceptible to fatigue. An overview is given from the reports which have dealt with either short-term or long-lasting SSCs in human subjects.

Short-Term SSC Fatigue

A special sledge apparatus [Kaneko et al., 1984; Aura and Komi, 1986a, b] has been developed to study isolated forms of eccentric and concentric exercises as well as their combinations (SSC). Gollhofer et al. [1987a, b] utilized this apparatus to study fatigue with normal healthy men, who performed 100 submaximal SSCs so that they were lying on the sledge with their heads towards the force plate, which was attached to the lower end of the sledge and perpendicular to the sliding metal bars. The exercise was then performed with both arms. Their results showed that during 100 SSCs the fatigue was characterized by increases in the contact times for both eccentric and concentric phases of the cycle (fig. 1). A more detailed graphic analysis revealed that the force-time curves during contact on the platform were influenced by fatigue. The initial force peak became higher and the subsequent initial drop of force more pronounced. More interestingly, however, the reflex contribution to sustain the repeated stretch loads became enhanced, especially when measured during the maximal drop-test condition before and immediately after the fatigue loading (fig. 2). Thus, in a non-fatigue state the muscles were able to damp the impact in SSC by a smooth force increase and by a smooth joint motion. However, repeated damping movements followed by the concentric action may have eventually become so fatiguing that the neuromuscular system changed its 'stiffness' regulation. This change was characterized especially by a high impact force peak followed by a rapid temporary force decline.

Stretch reflex contribution during fatigue could be interpreted to imply attempts of the nervous system to compensate, by increasing activation, the loss of the muscles' contractile force to resist repeated impact loads. It is, however, not known what could trigger the enhanced reflex contribution. One possible candidate could be accumulation of the metabolic products which induce acidosis in the milieu surrounding Ia-afferent nerve terminals. This hypothesis follows the observation of Fujitsuka et al. [1979]

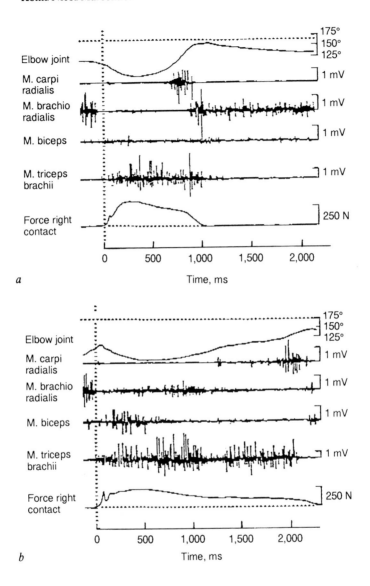

Fig. 1. Influence of exhaustive SSC exercises using the arms on EMG records and force-time curves. The subject performed 100 repeated SSCs on a special sledge apparatus, and was lying with the head towards the force plate. Note the increased period of hand contact on the force plate between the first *(a)* and the last *(b)* of the SSCs. Similarly, the initial force peak during contact increased with increasing fatigue. From Gollhofer et al. [1987b].

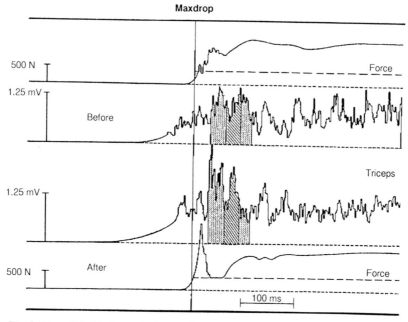

Fig. 2. The subject in figure 1 also performed maximal SSC (high drops against the force plate) before *(a)* and immediately after *(b)* 100 exhaustive SSCs. The initial force peaks and subsequent drops increased dramatically due to fatigue. Rectified EMG records suggest augmentation of the short and medium latency reflex components (the first two shaded areas in the EMG records) during the test immediately following the fatiguing SSC exercise. From Gollhofer et al. [1987b].

who recorded modulated Ia-afferent discharges in stretched muscle spindles of isolated frog muscle by quite small changes in extracellular pH value. The frequency of the Ia-afferent discharges was increased up to twice the value for normal pH conditions where the pH value in extracellular medium was lowered by 0.1–0.2. Whatever the reason, the fact remains that SSC exercises demonstrate a challenge to all interested in the fatigue situation where the various neural and mechanical components of stiffness regulation are also put under repeated stress. In the fatigue situation of SSC with isolated muscle preparations, the prestretch phase (eccentric) action has been suggested to act as protection against fatigue and thus delaying its onset [de Haan et al., 1991]. While this hypothesis seems attractive, it must be emphasized that the isolated muscle preparation with constant electrical stimulation differs from the real SSC, where neural activation including the reflectory influences are left to operate naturally.

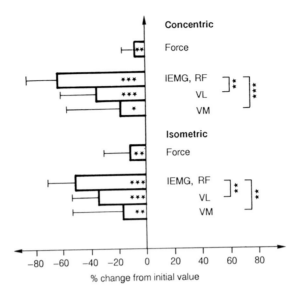

Fig. 3. The effects of 85-km X-C ski marathon on maximal concentric and isometric knee extension forces and EMG integrals. Measurements were made before (initial) and after the race. RF = m. rectus femoris; VL = m. vastus lateralis; VM = m. vastus medialis. * p<0.05; ** p<0.01; *** p<0.001. From Viitasalo et al. [1982].

Long-Lasting SSC Fatigue

Both cross-country (X-C) skiing [Viitasalo et al., 1982] and marathon running [Sherman et al., 1984; Komi et al., 1986; Nicol et al., 1991a, b] have been used as models to examine what effects the long-lasting SSC exercises have on the force production capacity and/or EMG activation of the leg extensor muscles. X-C skiing – especially that performed with the traditional-type diagonal technique – has been shown to demonstrate a chain of SSCs [Komi and Norman, 1987]. Viitasalo et al. [1982] studied 23 skiers who participated in the Wasa X-C ski marathon with the finishing times ranging from 5.37 to 9.49 h. The measurements taken the day before and 1–2 h after the race, demonstrated clear changes in maximal knee extension performance of concentric as well as isometric (fig. 3) actions. The maximal torque decreased significantly, but its reduction was much less than that of the maximal rate of torque production as well as its relaxation. As expected, the maximal IEMG levels also decreased significantly, and even more than the

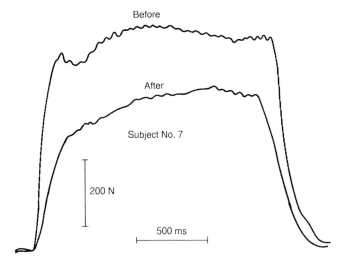

Fig. 4. Examples of the isometric knee extension torque-time curves measured for one individual skier before and after the 85-km Wasa X-C ski race. Unpublished material from the study of Viitasalo et al. [1982].

maximal torque. The fact that the rate of force development and its relaxation decreased more than the maximal torque level itself may have important relevance to the fatigue phenomenon. The examples of the torque-time curves before and after the ski marathon (fig. 4) demonstrate similar fatigue response as the motor unit mechanograms taken during the various phases of the progressing fatigue [Steg, 1964; Gydikov et al., 1976].

The leg extensor muscles in X-C skiing are subjected to smoother impact and stretch loads as compared to running. In running the duration of the phase including the initial impact peak and the subsequent braking (eccentric phase) is very short (50–120 ms) and the repeated loading will consequently have greater stretch-induced effects on the stiffness regulation than in X-C skiing. Thus, it would not be surprising to observe that a marathon run which is usually shorter in duration than a 75–90 km X-C ski race could induce more dramatic reductions in the force production. Maximal isometric knee extension has been shown to decrease during marathon on average by 26% [Nicol et al., 1991a] or 35.5% [Sherman et al., 1984]. As in X-C skiing the marathon run was associated with a significant change in the shape of the torque-time curve indicating that when the race and fatigue progressed the subjects needed more time to reach the certain absolute and relative force levels [Nicol et al., 1991a]. In the later study the overall performance reduction as judged by the maximal sprint tests started to develop after 20 km and went down by 15.7 ± 10%, the value which is similar to that observed a few years earlier [Komi et al., 1986].

Force platform, kinematic and electrogoniometer techniques can be used to study what changes take place in the SSC itself when measured either during running or jumping. In this regard a simple drop jump performed from a 50-cm height on the force platform is very sensitive to fatigue during a marathon run [Komi et al., 1986; Gollhofer et al., 1989]. The vertical jumping performance naturally decreased, but this decrease was associated with an increase in the braking (eccentric) contact time on the force plate. This clearly implies reduction in tolerance to and utilization of the stretch loads. This phenomenon has not been shown so clearly in the ground reaction force curves of the contact phase in running as was the case in the sledge experiments with the arm muscles (fig. 2), but the trend seems to be the same. Increase in the impact peak with subsequent rapid decline have been observed to occur in individual cases when the test was performed with the speed close to that of the marathon race (fig. 5). Nicol et al. [1991a] used similar techniques and were able to show fast drops in the vertical ground reaction force curves observed after initial impact both in the maximal sprint and five jump tests. Thus, occurrence of high initial force peaks followed by a rapid force decline suggests a similar fatigue mechanism as in the short-term SSC situations [Gollhofer et al., 1987]. When fatigue progresses the force becomes less absorbed by the tendomuscular system. The final result would be changes in the muscle stiffness characteristics and, hence, in the amount of storage of elastic energy in the braking (eccentric) phase of SSC. An impairment of good recoil is then expected to occur in the subsequent push-off phase. The ground reaction force curves taken during the latter half of the marathon race demonstrated a decrease in the average horizontal force during the push-off phase [Nicol et al., 1991a]. Concomitantly, both braking and push-off phase durations of the contact were longer as the fatigue progressed. Thus, it is likely that while the repeated stretch loads caused reduction in tolerance to stretch with subsequent increase in the braking time, the transfer from stretch to shortening also became longer. Reduced power during the eccentric phase would imply lesser storage of elastic energy [Bobbert et al., 1990]. Increased coupling time between stretch and shortening will in turn suggest further loss in the potential energy by dissipation as heat [Cavagna et al., 1968]. The final influence would then be reduced potentiation of performance in the SSC. The longer push-off phase in the

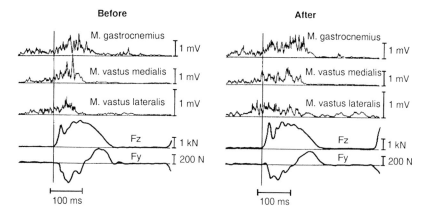

Fig. 5. Examples of the vertical (Fz) and horizontal (Fy) ground reaction forces as well as of EMG activities of the selected muscles from a ground contact phase of running at constant submaximal speed before and after the marathon. From Komi et al. [1986].

sprint performance test suggests an unsuccessful attempt of the neuromuscular system to compensate for the lost elastic potential.

Experiments with marathon running have shown 36–42% reduction in maximal IEMG activity of the leg extensor muscles in the knee extension test [Nicol et al., 1991b]. When the recordings were made in running at constant marathon speed, all the investigated muscles (vastus medialis, vastus lateralis and gastrocnemius) increased their activity especially in the push-off phase [Komi et al., 1986] (fig. 2). In addition, EMG/force ratio increased almost twofold for the push-off phase during the course of the marathon. Despite maintained high activation during the preactivation and braking phases, the same constant running velocity (=same propelling forces) could not be performed without this dramatic increase in the EMG activation. This phenomenon can be regarded as an attempt of the neuromuscular system to compensate for and match the reduced contractile and recoil characteristics of the leg extensor muscles. Increased EMG activity must also be regarded as indirect evidence for increased energy expenditure. EMG and $\dot{V}O_2$ have been shown to be well interrelated and change in parallel in the course of the exercise [Bigland-Ritchie et al., 1979]. Increased energy expenditure should then imply reduced mechanical efficiency of running, a logical consequence of fatigue.

References

Aura, O.; Komi, P.V.: Mechanical efficiency of pure positive and pure negative work with special reference to the work intensity. Int. J. Sports Med. *7:* 44–49 (1986a).

Aura, O.; Komi, P.V.: Effects of prestretch intensity on mechanical efficiency of positive work and elastic behaviour of skeletal muscle in stretch-shortening cycle exercise. Int. J. Sports Med. *7:* 137–143 (1986b).

Bigland-Ritchie, B.; Jones, D.A.; Woods, J.J.: Excitation frequency and muscle fatigue. Electrical responses during human voluntary and stimulated contractions. Expl. Neurol. *64:* 414–427 (1979).

Bobbert, M.F.: Drop jumping as a training method for jumping ability. Sports Med. *9:* 7–22 (1990).

Cavagna, G.A.; Dusman, B.; Margaria, R.: Positive work done by the previously stretched muscle. J. appl. Physiol. *24:* 21–32 (1968).

De Haan, A.; Lodder, M.A.N.; Sargeant, A.T.: Influence of an active pre-stretch on fatigue of skeletal muscle. Eur. J. appl. Physiol. *62:* 268–273 (1991).

Fujitsuka, N.; Ohkuwa, T.; Mitsui, J.; Utsumo, T.; Ozawa, N.: Intraindividual variation of oxygen debt as related to menstrual cycle after submaximal bicycle exercise. Bull. Nagoya Inst. Technol. *31:* 427–432 (1979).

Gollhofer, A.; Komi, P.V.; Fujitsuka, N.; Miyashita, M.: Fatigue during stretch-shortening cycle exercises. II. Changes in neuromuscular activation patterns of human skeletal muscle. Int. J. Sports Med. *8:* suppl., pp. 38–47 (1987b).

Gollhofer, A.; Komi, P.V.; Hyvärinen, T.: Auswirkungen eines Marathonlaufes auf die Leistungscharakteristik und das Innervationsverhalten der Beinstreckmuskulatur. Dt. Z. Sportmed. *40:* 348–354 (1989).

Gollhofer, A.; Komi, P.V.; Miyashita, M.; Aura, O.: Fatigue during stretch-shortening cycle exercise. I. Changes in mechanical performance of human skeletal muscle. Int. J. Sports Med. *8:* suppl., pp. 71–78 (1987a).

Gregor, R.J.; Komi, P.V.; Järvinen, M.: Achilles tendon forces during cycling. Int. J. Sports Med. *8:* suppl., pp. 9–14 (1987).

Gydikov, A.; Dimitrov, G.; Kosarov, D.; Dimitrova, N.: Functional differentiation of motor units in human opponens pollicis muscle. Experimental Neurology *50:* 36–47 (1976).

Hoffer, J.A.; Andreassen, S.: Regulation of soleus stiffness in premammillary cats. Intrinsic and reflex components. J. Neurophysiol. *45:* 267–285 (1981).

Huijing, P.A.: Elastic potential of muscle; in Komi, PV: Strength and power in sport (Blackwell Scientific Publications, Oxford, 1991).

Kaneko, M.; Komi, P.V.; Aura, O.: Mechanical efficiency of concentric and eccentric exercises performed with medium to fast contraction rates, Scand. J. Sport Sci. *6:* 15–20 (1984).

Komi, P.V.: Physiological and biomechanical correlates of muscle function. Effects of muscle structure and stretch-shortening cycle on force and speed; in Trejung: Exercise and sport science reviews, vol. 12, pp. 81–121 (Collamore Press, Lexington 1984).

Komi, P.V.: Relevance of in vivo force measurements to human biomechanics. J. Biomechanics *23:* suppl., pp. 23–34 (1990).

Komi, P.V.; Hyvärinen, T.; Gollhofer, A.; Mero, A.: Man-shoe-surface interaction. Special problems during marathon running. Acta Univ. Oulu *179:* 69–72 (1986).

Komi, P.V.; Norman, R.W.: Preloading of the thrust phase in cross-country skiing. Int. J. Sports Med. *8:* suppl., pp. 48–54 (1987).

Nicol, C.; Komi, P.V.; Marconnet, P.: Fatigue effects of marathon running on neuromuscular performance. I. Changes in muscle force and stiffness characteristics. Scand. J. med. Sci. Sports *1:* 10–17 (1991a).

Nicol, C.; Komi, P.V.; Marconnet, P.: Fatigue effects of marathon running on neuromuscular performance. II. Changes in force, integrated electromyographic activity and endurance capacity. Scand. J. med. Sci. Sports *1:* 18–24 (1991b).

Sherman, W.M.; Armstrong, L.E.; Murray, T.M.; Hagerman, F.C.; Costill, D.L.; Staron, R.C.; Ivy, J.L.: Effect of a 42.4 km footrace and subsequent rest or exercise on muscular strength and work capacity. J. appl. Physiol. *57:* 1668–1673 (1984).

Steg, G.: Efferent muscle innervation and rigidity. Acta. physiol. scand. *61:* 5 (1964).

Viitasalo, J.T.; Komi, P.V.; Jacobs, I.; Karlsson, J.: Effects of a prolonged cross-country skiing on neuromuscular performance; in Komi, Exercise and sport biology (International series on sport sciences), vol. 12, pp. 191–198. Champaign, Human Kinetics Publishers, 1982.

Prof. Paavo V. Komi, PhD, Department of Biology and Physical Activity, University of Jyväskylä, SF–40100 Jyväskylä (Finland)

Marconnet P, Komi PV, Saltin B, Sejersted OM (eds): Muscle Fatigue Mechanisms in
Exercise and Training. Med Sport Sci. Basel, Karger, 1992, vol 34, pp 182–194

Heart Muscle Metabolism and Function during Heavy Exercise

Lennart Kaijser

Karolinska Institute, Department of Clinical Physiology, Huddinge Hospital,
Stockholm, Sweden

Dynamic physical exercise with large muscle groups demands a substantial increase in cardiac output. Thus, a healthy young man of average physical fitness increases his cardiac output from 5–10 liters/min at rest to about 25 liters/min when exercising at an intensity corresponding to his maximal oxygen uptake of about 3.5 liters/min. Catheterization studies have shown that on transition from rest to exercise the stroke volume is increased by about 30%. The stroke volume then remains more or less unaltered with increased load, the increase in cardiac output being produced by increased heart rate [1]. This means that the stroke volume is a crucial characteristic of the capacity of the central circulation to provide the exercising musculature with oxygenated blood.

A number of questions may be raised with regard to the ability of the heart to maintain adequate pump function, e.g. what makes it possible for the top rank athlete to reach twice the oxygen uptake of our average healthy young man; is the stroke volume maintained even during exhaustive exercise; is the substrate supply of the heart and the coronary circulation enough under all conditions of exercise? These questions will be addressed in the following pages.

Heart Size

The large maximal oxygen uptake capacity of the athlete would presuppose a large stroke volume which in turn would postulate a large heart

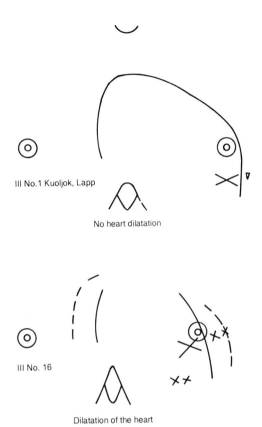

III No.1 Kuoljok, Lapp

No heart dilatation

III No. 16

Dilatation of the heart

Fig. 1. The heart borders, as defined by percussion, show a large heart in the Lapp Kuoljok, winner of the Falun-Gävle 'ultramarathon' ski race. The less successful competitor, who finished number 16 in the Falun-Gävle race shows at the start of the race (full-line) smaller heart size than the winning Lapp and a 'dilatation' of the heart at the finish (broken line).

volume. Already in the 18th century it was observed that physically active animals had larger hearts in relation to their body weight than more sedentary ones [2]. About a hundred years ago, Henschen [3] studied the question of the heart size in athletes. Using simple methods of physical examination of the chest, defining the borders of the heart by percussion, he studied heart sizes of participants in cross-country ski races on the college and national elite level. He found that the most successful participants had the largest heart volumes (fig. 1, 2). From this he drew the conclusion that a

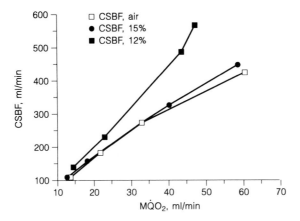

Fig. 2. Coronary sinus blood flow (CSBF) as a function of myocardial oxygen consumption at rest and during successively heavier exercise up to fatigue breathing air, 15% O_2 and 12% O_2, respectively (simulating 2,500 and 4,500 m above sea level).

large heart volume was a prerequisite for a high capacity to perform endurance exercise. The question was also raised whether this was a normal and adequate adaptation to an increased demand or a pathological consequence of too much strain. To analyze if the large hearts were better suited to take the strain of extreme exercise he measured the heart volume before and after an 'ultramarathon' ski race over 95 km in woods and over mountains without prepared tracks. He found that the most successful participants were not only characterized by larger hearts, but also that their heart volumes did not increase during the race, whereas the less successful participants had smaller heart volumes which, however, became larger after than before the race (fig. 2). He interpreted the last-mentioned finding as a dilatation of the heart, which he considered as a sign of insufficient ability to maintain contractile function under conditions of extreme strain.

Later, with the techniques of heart X-ray and cardiovascular catheterization available, more precise measurements of heart volume could be done together with measurements of the stroke volume. These studies verified that endurance athletes had large heart volumes and that in healthy subjects – including athletes as well as sedentary subjects – there was a correlation between the heart volume on the one hand and on the other the stroke volume, the total hemoglobin amount and the blood volume (which can be taken as a measure of the size of the vascular system) and the physical working capacity [4]. Furthermore, longitudinal training studies showed parallel increases in heart volume, blood volume and physical working capacity [5]. The findings were taken to support the contention that the large heart of the athlete represented an adequate adaptation, since in patients with hearts enlarged by heart disease the enlargement was greater than expected from the blood volume or the physical working capacity [4]. Recent studies, utilizing echocardiography, have shown that the enlargement of the heart in endurance-trained subjects consists of proportional increases in internal diameters of heart chambers and left ventricular wall thickness, i.e. an adequate adaptation to produce a large stroke volume, suggesting no adverse effect on the heart muscle [6, 7]. A few studies have suggested that training characterized by a significant proportion of isometric muscle contraction, which is accompanied by a pronounced increase in cardiac afterload, affects the heart differently than endurance training, increasing the left ventricular wall thickness more than the ventricular cavity diameter [8]. It has also been discussed whether or not such an adaptation is 'bad' for the heart. However, neither the finding nor the interpretation of it is as yet firmly established [6].

It can be concluded that present evidence indicates that endurance athletes have larger than normal hearts, that the heart enlargement comprises chamber dimensions and wall thickness to a proportional extent, is paralleled by increases in stroke volume, blood volume and physical working capacity and that it is not related to any known negative effect on the heart muscle.

Cardiac Function during Exhaustive Exercise

After an increase in stroke volume on transition from rest to exercise the stroke volume remains approximately constant irrespective of work intensity during exercise of moderate duration [1]. It has been discussed if it is possible to maintain the stroke volume up to maximal exercise or if maximal work is paralleled by a decrease in stroke volume. Holmgren and Ovenfors [9] followed the heart volume during exercise by bi-plane serial X-ray, ECG-triggered for exposure in end diastole. When the subjects exercised in the sitting position on the cycle ergometer the heart volume increased on transition from rest to exercise, but decreased gradually, although only slightly, when exercise intensity was increased to close to maximal. The finding could be explained by either a reduced stroke volume during heavy exercise or a parallel reduction in both end-systolic and end-diastolic volume with a maintained stroke volume. A few subjects repeated the exercise in the supine position. At rest as well as at all work loads the heart volume was larger in the supine than in the sitting position, but also in the supine position a slight decrease occurred at the highest loads. This strongly suggests that the decrease in heart volume at high work load was caused by a redistribution of blood to the periphery in combination with a more complete emptying during systole rather than by a reduced myocardial function. Catheterization studies during exercise at stepwise increased load suggest a maintained stroke volume during maximal exercise [10]. Thus, most evidence suggests that short-term maximal exercise is not limited by myocardial power.

During prolonged submaximal exercise of 60 min duration there is a continuous increase in heart rate, both if exercise is executed in the sitting and in the supine position [11, 12]. The cardiac output remains constant, implying a gradual decrease in stroke volume. Since the central venous pressure gradually falls, the probable explanation is a redistribution of blood to the periphery with reduced filling of the heart rather than a reduced cardiac function. However, in experiments where prolonged exercise was continued to exhaustion and heart volume measured, triggered by the ECG, in end-diastole, the heart volume of half the subjects increased during the last 10 min before exhaustion, after having gradually decreased during the initial 40–50 min of exercise [13]. The findings were the same in supine as in sitting bicycle exercise. It leaves the possibility open that cardiac function is altered on approaching exhaustion during prolonged exercise, although the possibility also remains that an extreme increase in sympathetic activity facilitates a redistribution of the blood volume in a central direction.

Recent studies, utilizing echocardiography, have added indications that heavy exercise of extremely long duration could lead to 'cardiac fatigue'. Thus, studies in connection with the Hawaii Ironman Triathlon (long distance swimming + cycling + running) showed a reduced fractional shortening of the left ventricular wall immediately after, compared to before, the competition [14]. The decreased fractional shortening was recorded together with an unaltered arterial pressure but reduced ventricular cavity dimension. Thus, it cannot be excluded that it was related to a reduced preload. Against this explanation is the fact that the fractional shortening but not the cavity dimension had returned to normal 1 day after the race, i.e. the race produced an altered relationship between fractional shortening and preload.

Of interest in this connection is the effect of prolonged psychological stress on the heart. In a laboratory experiment where healthy young men were subjected to extreme psychological stress together with sleep deprivation continuously over 3 days and nights, flat or negative T waves developed in

15% of the participants. The ECG changes remained for 12–36 h, i.e. longer than the elevations in heart rate or urinary catecholamine excretion [Kaijser and Levi, unpubl, observation].

Coronary Circulation

The myocardial metabolism is normally fully aerobic. During dynamic exercise at an intensity corresponding to the maximal oxygen uptake of the individual the myocardial oxygen consumption is increased about 5-fold compared to at rest [15]. This increase is covered mainly by an augmented coronary blood flow (4-fold increase) while a slightly more complete oxygen extraction contributes to a lesser extent. Even during maximal physical exercise, the oxygen saturation in the vein draining the myocardium, the coronary sinus, is 20–25%, i.e. far higher than in the blood draining the maximally exercising leg musculature [15]. Therefore, it would seem that the heart muscle has more than enough oxygen supply. However, the coronary sinus blood represents a mixture of the myocardial drainage, and conditions for perfusion as well as oxygen consumption may diverge between different parts of the myocardium. Thus, the subendocardial layer of the left ventricle seems to have a greater perfusion per unit weight than the subepicardial layer and the right ventricle. However, the perfusion in relation to the oxygen demand may be smaller subendocardially than subepicardially [16]. Furthermore, the perfusion of the subendocardium is probably almost totally confined to diastole, whereas in other parts a significant perfusion takes place also during systole. Consequently, there is the possibility of an insufficient blood supply to parts of the heart muscle, especially under conditions of high cardiac power output and short relative duration of diastole. To what extent such conditions prevail during any mode of exercise in the healthy individual is not known.

Under aerobic conditions, lactate is taken up by the heart muscle and used as substrate for the oxidative metabolism. This is mirrored in a positive arterial-coronary sinus difference of lactate, which is found under all modes and intensities of exercise in healthy men [17, 18]. However, uneven blood flow distribution with limited areas of anaerobic metabolism would escape detection by the chemical measurement of arterial-coronary sinus differences. Intravenous infusion of ^{14}C-labeled lactate may be used to detect small amounts of lactate released under conditions of myocardial net uptake of lactate, a sign of smaller myocardial areas with anaerobic metabolism at the same time as the major part of the heart muscle is aerobic [19]. However, in our own studies we have not found indications of such a lactate release in healthy young men during heavy dynamic exercise [15].

In an attempt to establish whether or not there is 'a coronary flow reserve' in excess of the flow during maximal physical exercise, we have studied healthy young men by coronary sinus catheterization during exercise up to a maximal level under conditions of varying degrees of hypoxia [15]. The studies have shown that it is possible to compensate for a reduction to 90% in arterial oxygen saturation. The compensation is at low cardiac power output, mainly brought about by a more complete oxygen extraction resulting in a lowered coronary sinus oxygen saturation, leaving the arterial-coronary sinus oxygen difference insignificantly reduced and the coronary blood flow insignificantly increased. With increased cardiac power output during exercise of successively increased intensity, the compensatory mechanism becomes an increased coronary blood flow with no difference in coronary sinus oxygen saturation compared to air breathing. During maximal exercise in hypoxemia the coronary blood flow can be increased by about 25% above that prevailing during maximal dynamic exercise breathing air (fig. 2). As a result a reduction in arterial oxygen content by about 10%, corresponding to an altitude of about 2,500 m is tolerated without a detectable decrease in myocardial oxygen uptake. Yet, under hypoxemic conditions ^{14}C-lactate data indicate some myocardial lactate release already at a work intensity below that which requires maximal coronary blood flow, suggesting that part of the heart is not sufficiently supplied with oxygen.

Myocardial Substrate Metabolism

At rest in the basal state, i.e. in the morning after an overnight fast, the quantitatively most important substrate is plasma free fatty acids (FFA), which cover about 50% of the substrate oxidation [17]. In addition to this, plasma triglycerides cover about 15%. Thus, fat accounts for 2/3 of the oxidative metabolism, a figure which is supported by measurements of the respiratory quotient over the heart. The remaining 1/3 is covered by carbohydrates, mainly glucose and lactate, covering about 20 and 5–10% of the oxidative metabolism, respectively. Other substrates, e.g. ketone bodies, are of minor quantitative importance in the basal state. Adding together the contributions of the mentioned substrates, we find that they cover 100% of the oxidative metabolism, i.e. blood-borne substrates supply the whole need

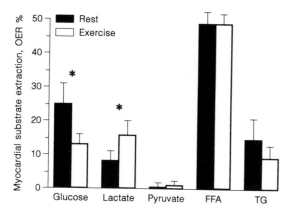

Fig. 3. Myocardial oxidative substrate metabolism in healthy young men at rest and during prolonged exercise at 40–50% of maximal O$_2$ uptake. Substrate utilization is expressed as the fraction of the total oxidative metabolism which each substrate would cover if fully oxidized (OER, %).

(fig. 3). This is of course essential, since unlike the skeletal muscle, the heart has no resting periods during which it can replenish its endogenous substrate stores.

At least two factors may affect the conditions for substrate utilization during exercise: the increased cardiac power output and altered substrate supply by changing blood concentrations. A moderate increase in cardiac power output seems not to alter the proportions of substrates utilized by the heart muscle. This may be exemplified by the conditions during atrial pacing (fig. 4). Actual physical exercise, on the other hand, is almost always accompanied by alterations in concentrations of substrates in the blood, which would affect the myocardial substrate extraction. Three factors related to substrate and hormone concentrations can be regarded as major determinants of myocardial extraction of a given substrate: its concentration in arterial blood, the interaction between metabolism of different substrates and the concentrations of some hormones, notably insulin [20]. The importance of the concentration in arterial blood may be exemplified by the relationship between extraction by the heart and arterial concentration of plasma FFA. In a study where the lowest FFA concentrations were achieved by inhibiting the lipolysis in adipose tissue by nicotinic acid, a linear relationship was found over a wide range (fig. 5). It was not firmly established

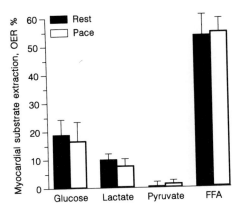

Fig. 4. Myocardial oxidative substrate metabolism at rest and during atrial pacing at 130–140 beats/min.

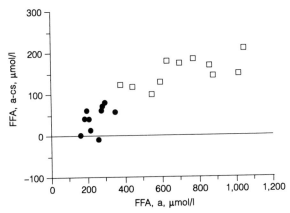

Fig. 5. Myocardial extraction of plasma free fatty acids (FFA) as a function of arterial plasma FFA concentration in healthy young men. The filled symbols refer to the conditions of lowered plasma FFA concentration achieved by nicotinic acid.

if there is a lower limit below which no translocation of FFA into the myocardial cells can take place and if an upper limit for the translocation capacity of FFA into myocardial cells exists.

The second important determinant of myocardial substrate uptake is the interaction between utilization of different substrates, most importantly the interaction between carbohydrate and fat metabolism. In the 1960s, Randle et al. [21], working with isolated rat heart preparations, proposed the existence of a 'glucose fatty acid cycle' to the effect that an increased utilization of fat would inhibit the utilization of carbohydrate. The proposed mechanism was an increased production of acetyl-CoA from the β-oxidation of long-chain fatty acids inhibiting the pyruvate dehydrogenase complex, thereby slowing down the oxidative utilization of pyruvate. Furthermore, fatty acid oxidation produces an increased amount of citrate which is an inhibitor of phosphofructokinase, considered to be a rate-limiting enzyme in the glycolysis. The importance of the glucose-fatty acid cycle under physiological conditions was questioned until we could show in man, utilizing nicotinic acid to decrease the plasma FFA concentration, that a decrease in FFA concentration and myocardial extraction was accompanied by a quantitatively related increase in myocardial extraction of glucose, lactate and pyruvate [22].

In addition to the two mentioned factors of importance for the myocardial substrate choice insulin promotes the myocardial extraction of glucose.

During exercise of submaximal intensity the arterial glucose concentration remains constant and the lactate concentration is moderately elevated. In this situation the myocardial utilization of different substrates is not much different from that at rest. Myocardial RQ is unaltered indicating unaltered proportions of fat/carbohydrate utilization. A slightly greater extraction of lactate is related to the slightly higher blood concentration of lactate.

During heavy exercise, at or above an intensity corresponding to the maximal oxygen uptake of the subject, the availability of lactate in the blood is great. In this situation lactate is the dominating substrate for the myocardium and the uptake of lactate may well be enough to cover the whole oxidative substrate utilization.

While during heavy, short-term exhaustive exercise the supply of blood-borne substrates seem to be enough for the heart, the conditions during prolonged exercise may be divergent. Thus, after 120 min of exercise at 40–50% of maximal oxygen uptake the myocardial extraction of FFA is similar to that at rest. This is below the extraction expected from the arterial FFA concentration, which has increased to about twice the resting level [17]. Although the total extraction of substrates is not significantly smaller than the total substrate oxidation (as estimated from the myocardial oxygen extraction; fig. 3) there is a significant myocardial release of glycerol, which is not seen at rest or during work of shorter duration (fig. 6). The glycerol release may signify an intramyocardial lipolysis, i.e. a utilization of triglycerides

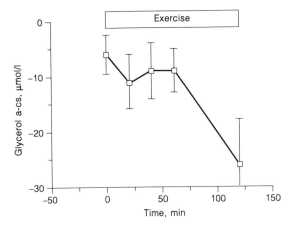

Fig. 6. Myocardial glycerol release at rest and at intervals during 120 min exercise at 40–50% of maximal oxygen uptake.

stored in the myocardial cells. The utilization of intramyocardial triglyceride stores may be secondary to an insufficient extraction of plasma FFA. Again, the question may be raised whether there is an upper limit for FFA uptake. Alternatively, the smaller than expected extraction of plasma FFA may be secondary to an increased intramyocardial lipolysis.

Thus there are indications that, unlike in all other situations studied, the healthy heart muscle, during exercise of extreme duration, starts to utilize endogenously stored substrates. If this has a negative effect on the cardiac function is not known. Nor is it known if the cardiac function is affected by the kind of substrate utilized. The combustion of fat requires more oxygen than the combustion of carbohydrates, which is deleterious in conditions of impaired blood or oxygen supply, e.g. in patients with ischemic heart disease [23]. In ischemic heart disease the great intramyocardial accumulation of fat may also have negative effects not related to the increased oxygen need. Thus it seems to provoke arrythmias [24]. If a high rate of utilization of FFA has a negative effect on the function of the well-oxygenated healthy heart during conditions of heavy exercise is not known, however, and the possible relationship between metabolic alterations and cardiac function [14] during extreme prolonged exercise remains to be elucidated.

References

1 Bevegard S, Holmgren A, Jonsson B: Circulatory studies in well trained athletes at rest and during heavy exercise with special reference to stroke volume and the influence of body position. Acta Physiol Scand 1963;57:26–50.

2 Robinson S: A dissertation on the food and discharges of human bodies. London, Nourse 1748, pp 1–120 (at the Lamb against Katherine Street in the Strand).

3 Henschen SE: Skidlauf und Skidwettlauf. Eine medizinische Sportsstudie; in von Fischer G (ed): Mitteilungen aus der Medizinischen Klinik zu Upsala. Jena, 1899, pp 1–74.

4 Sjöstrand T: Volume and distribution of blood and their significance in regulating the circulation. Phys Rev 1953;33:202–217.

5 Holmgren A, Mossfeldt F, Sjöstrand T, Ström G: Effect of training on work capacity, total hemoglobin, blood volume, heart volume and pulse rate in recumbent and upright positions. Acta Physiol Scand 1960;50:72–83.

6 Oakley D: Cardiac hypertrophy in athletes. Br Heart J 1984;52:121–123.

7 Shapiro IM: Physiological left ventricular hypertrophy. Br Heart J 1984;52:130–135.

8 Fagard R, Aubert A, Staessen J, Eynde EV, Vanhees L, Amery A: Cardiac structure and function in cyclists and runners. Comparative echocardiographic study. Br Heart J 1984;52:124–129.

9 Holmgren A, Ovenfors CO: Heart volume at rest and during muscular work in the supine and in the sitting position. Acta Med Scand 1960;167:267–277.

10 Ekblom B, Hermansen L: Cardiac output in athletes. J Appl Physiol 1968;25:619–625.

11 Ekelund L-G: Circulatory and respiratory adaptation during prolonged exercise of moderate intensity in the sitting position. Acta Physiol Scand 1967;69:327–340.

12 Ekelund L-G: Circulatory and respiratory adaptation during prolonged exercise in the supine position. Acta Physiol Scand 1966;68:382–396.

13 Ekelund L-G, Holmgren A, Ovenfors CO: Heart volume during prolonged exercise in the supine and sitting position. Acta Physiol Scand 1967;70:88–98.

14 Douglas PS, O'Toole ML, Hiller WDB, Hackney K, Reichek N: Cardiac fatigue after prolonged exercise. Circulation 1987;76:1206–1213.

15 Kaijser L, Grubbström J, Berglund B: Coronary circulation in acute hypoxia. Clin Physiol 1990;10:259–263.

16 Hoffman JIE, Buckberg GD: The myocardial supply – demand ratio: a critical review. Am J Cardiol 1978;41:327–332.

17 Kaijser L, Lassers BW, Wahlqvist ML, Carlson LA: Myocardial lipid and carbohydrate metabolism in fasting men during prolonged exercise. J Appl Physiol 1972;32:847–858.

18 Keul J, Doll E, Steim H, Homburger H, Kern H, Reindell H: Über den Stoffwechsel des menschlichen Herzens. I. Die Substratversorgung des gesunden menschlichen Herzens in Ruhe, während und nach körperlicher Arbeit. Pflügers Arch Physiol 1965;282:1–16.

19 Gertz EW, Wisneski JA, Neese R, Bristow JD, Searle GL, Hanlou JT: Myocardial lactate metabolism: Evidence of lactate release during net chemical extraction in man. Circulation 1981;63:1273–1279.

20 Kaijser L: Regulatory mechanisms in the interaction of lipid and carbohydrate metabolism in cardiac and skeletal muscle in man; in Carlson LA, Pernow B (eds):

Metabolic Risk Factors in Ischemic Cardiovascular Disease. New York, Raven Press, 1986, pp 149–158.

21 Randle RJ, Garland PB, Hales CN, Newsholme EA: The glucose fatty acid cycle. Its role in insulin sensitivity and the metabolic disturbances of diabetes mellitus. Lancet 1963;i:785–789.

22 Lassers BW, Kaijser L, Wahlqvist ML, Carlson LA: Relationship in man between plasma free fatty acids and myocardial metabolism of carbohydrate substrates. Lancet 1971;ii:448–450.

23 Kaijser L, Carlson LA, Eklund B, Nye ER, Rössner S, Wahlqvist ML: Substrate uptake by the ischaemic human heart during angina induced by atrial pacing; in Oliver MF, Julian DG, Donald KW (eds): Effect of Acute Ischaemia on Myocardial Function. Edinburgh, Churchill Livingstone, 1972, pp 223–233.

24 Kurien VA, Yates PA, Oliver MF: The role of free fatty acids in the production of ventricular arrhythmias after coronary artery occlusion. Europ J Clin Invest 1971;1: 225–241.

L. Kaijser, MD, Department of Clinical Physiology, Huddinge Hospital,
S–141 86 Huddinge (Sweden)

Marconnet P, Komi PV, Saltin B, Sejersted OM (eds): Muscle Fatigue Mechanisms in Exercise and Training. Med Sport Sci. Basel, Karger, 1992, vol 34, pp 195–206

Exogenous Substrate Utilization during Prolonged Exercise: Studies with ^{13}C-Labeling

François Péronnet, Eudoxie Adopo, Denis Massicotte

Département d'éducation physique, Université de Montréal et Département de kinanthropologie, UQAM, Montréal, Qué., Canada

Introduction

Performance during long-duration exercise depends partly on the availability of energy substrates for muscle contraction. Carbohydrate (CHO) ingestion before (1-3 h) or during the exercise period is a common practice in order to attempt to improve substrate delivery to working muscle [1, 2]. The efficiency of such a practice depends on the amount of exogenous CHO ingested which is actually oxidized. The purpose of this chapter is to review the data concerning oxidation of exogenous substrates (mainly CHO) during prolonged exercise, which have been obtained mainly using ^{13}C-labeling, and to outline possible methodological problems associated with this technique.

Methodological Considerations

Principles of Tracer Studies

Quantitative data concerning oxidation rates of exogenous substrates during exercise arise from studies using ^{14}C or ^{13}C-labeled substrates. In these studies, the amount of exogenous substrate actually oxidized can be computed from the ratio 'tracer released under the form of labeled CO_2 (*C released)/tracer administered (*C administered)', which is equal to the ratio 'substrate oxidized/substrate administered': The *C recovered is equal to:

*C released = *C expired + *C trapped – *C background

where *C expired is the total amount of tracer emitted in the expired gas in the form of *CO_2; *C trapped is the amount of tracer which has been converted into *CO_2 when the substrate was oxidized, but which remains within the body at the end of the observation period; and *C background is the amount of tracer emitted in the form of *CO_2 but which arises from the oxidation of endogenous substrates and not from the oxidation of the exogenous substrate. When ^{14}C is used as tracer, the value of ^{14}C background can be considered as negligible, since the amount of ^{14}C which occurs naturally in the foodstuffs and thus into the endogenous substrate pools, is very small when compared to the amount of tracer administered and recovered. However, for ethical reasons, only three studies have been conducted using ^{14}C-labeled substrates [3–5]. Since this period only ^{13}C-labeled substrates have been used. Because ^{13}C is an abundant isotope which represents about 1.1% of the carbon element on earth, an appropriate correction should be made for ^{13}C background when ^{13}C release is measured.

*C Trapping in Bicarbonate Pools

Bicarbonate/carbonate pool in the body corresponds to about 130 liters of CO_2 including about 110 liters of CO_2 in the form of calcium carbonates in bone tissue [6]. The bone carbonate pool constitutes a very slow-exchanging pool (SEP) in which an amount of about 40 ml of CO_2 arising from decarboxylation of energy substrates disappears each minute. The mean residence time of a CO_2 molecule entering the SEP being very long (2–3 days), the SEP can be considered as a CO_2 sink where, at rest, approximately 20% of the CO_2 produced (40 ml/250 ml) is 'irreversibly' trapped. Accordingly, the recovery of *C arising from ingested or infused labeled bicarbonate or oxidation of exogenous labeled substrates, is about 80% at rest [7]. A correction, thus, should be made in order to compute *C released from *C expired [8]. However, as a result of exercise, CO_2 production increases and the amount of CO_2 lost into the SEP can diminish because of the reduction in bone blood flow. The recovery of *C arising from oxidation of exogenous labeled substrated can thus be considered as complete [6, 9]. The value of *C trapped corresponds only to the amount of *CO_2 produced during the exercise period, which has not been emitted into the atmosphere at the end of the observation period, and which remains diluted in the fast exchanging bicarbonate pool (i.e. sodium and potassium bicarbonates in the blood and soft tissues: about 20 liters of CO_2).

[13]C Background

Isotopic Fractionation. Carbon 13 is an abundant stable isotope of carbon: about 1.1 atoms of carbon in 100 being [13]C. The abundance of carbon 13 is expressed by the isotopic composition, which is the ratio $^{13}C/^{12}C$ (R) expressed in atoms per one hundred atoms (%) or in delta per mil (‰) by reference to a standard using Craig's formula [10, p. 485].

The abundance of ^{13}C in naturally occurring organic compounds varies due to isotopic fractionation along the biochemical pathways [10–14]. Isotopic fractionation is a reduction in ^{13}C abundance which occurs along a chemical pathway. This phenomenon tends to favor molecules with ^{12}C, which are less heavy and have a smaller binding energy than molecules containing [13]C. Isotopic fractionation explains why, along the food chain, beginning with inorganic CO_2, the isotopic ratio of organic compounds systematically diminishes. Moreover, lipids, which are derived from CHO (exclusively in plants, largely in animals), have a lower isotopic ratio than CHO. In plants, isotopic ratio depends on the isotopic ratio of the CO_2 used for photosynthesis and also depends on the type of photosynthetic cycle present in the plant. Indeed, the isotopic fractionation is larger in the classic Calvin cycle or C_3 cycle, which is present, for example, in wheat, potato, sugar beet, peanut, cotton or soya, than in the Hatch-Slack cycle or C_4 cycle, which is present for example in corn, cane-sugar, millet or sorghum [13, 15–17]. Isotopic composition of glycogen and fat in animals depends on the isotopic composition of plants which support their respective food chain, and, accordingly, are higher in animals ingesting food originating from C_4 photosynthetic plants [18]. In turn, the isotopic composition of CO_2 emitted by animals into the environment reflects the isotopic composition of their foodstuffs and of their food. This phenomenon is exemplified by the difference in isotopic composition of CO_2 expired by Western Europeans and North Americans. North Americans, and the animals they consume, have a diet which is more rich in cane-sugar and maize, than that of Western Europeans. Accordingly, the isotopic composition of their substrate pools and their expired CO_2 is higher than those of their counterparts from Western Europe [19].

Plants like corn and sugarcane, with a C_4 photosynthetic cycle which produces carbohydrates with a comparatively high isotopic ratio, provide inexpensive 'naturally' ^{13}C-labeled substrates, which can be used as tracers in metabolic studies in man [10].

Changes in ^{13}C Background. For the sake of estimating *C background, data concerning isotopic fractionation clearly indicate that ^{13}C is not distributed homogeneously in the various substrate pools of the body. Carbohydrates have a higher isotopic ratio than fats. In addition, one can suspect that glucose from the various carbohydrate pools, as well as free fatty acids from the various pools of fat, may also have different isotopic compositions. As a consequence, when the mixture of the various endogenous substrates oxidized is modified, the isotopic composition of CO_2 arising from this oxidation (i.e. ^{13}C background) will change. This phenomenon has been confirmed by several investigators. As early as 1970, Jacobson et al. [20] suggested that changes in ^{13}C in respiratory carbon dioxide from potato slices can be used as an indicator of the type of endogenous substrates oxidized. In 1971, Mosora et al. [21] reported that, in rats, a 24-hour fasting period, as well as administration of glucocorticoids or glucagon, resulted in changes in the isotopic composition of expired CO_2, reflecting changes in the mixture of endogenous substrates oxidized. In 1973, Duchesne et al. [22] recognized that the respective contributions of CHO, fat and proteins to energy metabolism determine the isotopic composition of expired CO_2. In the same year, Lacroix et al. [23, p. 446] also suggested that 'the rise in $^{13}C/^{12}C$ values (in expired CO_2) after oral glucose administration (...) might reflect not only the increased utilization of exogenous glucose but also *the reduced utilization of lipids,* which have a slightly lower ^{13}C content than carbohydrates and proteins'. As early as 1977, Schoeller et al. [19] reported that mild exercise increased the ^{13}C isotope abundance of exhaled CO_2. The authors attributed this phenomenon to the increased 'percentage of CO_2 arising in the skeletal muscle mass' and to change in 'the percentage of carbohydrate, protein and lipid being oxidized to CO_2' associated with exercise [pp. 418–419]. The effect of exercise on the isotopic composition of expired CO_2 was independently confirmed by Massicotte et al. [24] and by Wolfe et al. [25] in 1984. In subsequent studies from our laboratory [26–29], we have regularly reported changes in Rexp associated with changes in the mixture of endogenous substrates oxidized, in response to exercise without exogenous substrate ingestion. This phenomenon has recently been investigated by Barstow et al. [30] in various experimental conditions. Results from this study clearly indicate that, in response to exercise, Rexp regularly rises after a transient decrease in the first minutes of exercise.

It should be underlined that some authors have reported no change in Rexp from rest to exercise [31, 32]. These authors are working in Western Europe, and it can be hypothetized that the various endogenous substrate stores and pools of Western Europeans have more similar isotopic compositions than endogenous substrate stores and pools of North Americans. Accordingly, the question of ^{13}C background of expired CO_2 can be considered differently on the two continents.

The practical consequence of the heterogeneity of ^{13}C distribution in the various endogenous substrate stores and pools is that ^{13}C background (*C background) will be modified when the composition of the mixture of endogenous substrates is modified such as: (1) from rest to exercise (see above), and (2) during exercise when exogenous CHO are ingested. As discussed below, these changes in ^{13}C background have not been properly taken into account in the studies using ^{13}C-labeling.

Experimental Results

Studies with ^{13}C-Labeling

Since 1973, 18 studies of CHO oxidation using ^{13}C-labeling have been reported [26–28, 31–45]. These studies have been conducted in young healthy subjects (with the exception of the study by Krzentowski et al. [39] which includes a group of type I diabetic patients), with a wide range of working capacity, from sedentary subjects to highly trained recreational athletes. The exercise intensity used also varies widely both in terms of absolute $\dot{V}O_2$ (0.94–3.35 liters/min) or of %$\dot{V}O_2$ max (40–67%), and the exercise duration ranges from 50 to 240 min. Different types of CHO have been studied (glucose, fructose, sucrose, maltodextrins, glucose polymers and starch) with doses ranging from 50 up to 400 g, administered before or during the exercise period, in one single bolus or in several small fractionated doses, under the form of dilute solution of various concentrations (7–25%) or in solid form.

When exogenous CHO utilizations reported in all these studies are expressed as the oxidation rate, the values range between 0.15 to 0.84 g/min. As depicted in figure 1 the single most important factor which determines exogenous CHO oxidation rate is the absolute workload, expressed as $\dot{V}O_2$ in liters/min: when all the data points are taken into account, a good correlation exists between $\dot{V}O_2$ and exogenous CHO oxidation rate (r = 0.617). Interestingly, Pirnay et al. [41] have investigated, in the same study, the effect of various workloads on the oxidation rate of exogenous glucose. The four workloads chosen corresponded to 0.94, 1.67, 2.17 and 2.14 liters O_2/min and the exogenous glucose oxidation rates were 0.17, 0.36, 0.45 and 0.51 g/min, respectively. The correlation between the two sets of data is strong (r = 0.98) and the associated regression line (y = 0.014 + 0.190 $\dot{V}O_2$) is almost undistinguishable from the regression line computed from the 38 data points plotted on figure 1. It thus appears clear that the oxidation rate of

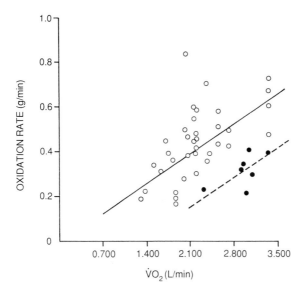

Fig. 1. Oxidation rate (g/min) of exogenous CHO plotted against $\dot{V}O_2$, in liters/min. Open symbols are data from literature ($r = 0.617$, oxidation rate $= 0.062 + 0.173 \cdot \dot{V}O_2$); closed circles are individual data from the present study ($r = 0.696$, oxidation rate $= -0.0204 + 0.182 \cdot \dot{V}O_2$).

CHO ingested before or during the exercise period is determined primarily by the power output sustained by the subjects.

Obviously, a wide dispersion of the data points around the regression line is present. Among several factors which can modify exogenous CHO oxidation for a given workload, the total amount of CHO ingested appears to play a role. For example, for $\dot{V}O_2$ values of around 2 liters/min, the oxidation rate reported by Jandrain et al. [37] for 50 g of glucose ingested is about 0.2 g/min. This rate increases to 0.39 g/min when 100 g of glucose are ingested [42], to 0.50 g/min when 200 g of glucose are ingested [40], and to 0.84 g/min when 400 g of glucose are ingested [40]. For a higher workload (3.35 liters O_2/min), Wagenmakers et al. [45] have shown that the oxidation rate of maltodextrins increases from 0.45 to 0.58, 0.65 and 0.70 g/min, when the amount administered increases, respectively, from 73 to 123, 174 and 248 g.

Other factors do not appear to modify systematically the oxidation rate of exogenous CHO in a great extent. For example, the oxidation rates reported are comparable when exogenous CHO are given (a) before the exercise period (in a single bolus); (b) during the exercise in a single bolus, or

(c) in several small fractionated doses. As for the concentration of the solution, this question was investigated in 1989 by Jandrain et al. [37] using three glucose solutions (50 g of glucose, in 200, 400 and 600 ml of water). The oxidation rates were essentially similar with the three solutions (0.18, 0.19 and 0.21 g/min). A limited number of studies have compared the oxidation rate of various CHO to the oxidation rate of exogenous glucose [26–28, 33, 35]. Results from Guézennec et al. [35] and from Massicotte et al. [27] indicate that the oxidation rates of glucose polymers and starch are not different from those of glucose in similar experimental conditions. On the other hand, the oxidation rate of exogenous fructose has been reported to be lower than that of glucose by Guézennec et al. [35] and by Massicotte et al. [26–28], but not by Décombaz et al. [33] (0.50 vs. 0.43 g/min). Results from Décombaz et al. [33] are difficult to interpret because fructose is, indeed, absorbed at a slower rate than glucose and has to be converted into glucose by the liver before being oxidized. Accordingly, its metabolic availability would be expected to be lower than that of exogenous glucose.

In order to complete this review of oxidation rates of exogenous CHO during prolonged exercise using ^{13}C-labeling, we should mention: (1) the experimental results reported by Gérard et al. [34] which indicate that exogenous sucrose utilization was reduced from 0.41 to 0.23 g/min, if the enzyme responsible for the hydrolysis of sucrose was inhibited by acarbose, and (2) the experimental result reported by Krzentowski et al. [39] which indicated that the oxidation rate of exogenous glucose was lower in diabetic patients deprived of insulin (0.19 vs. 0.41 g/min in control subjects) but was restored by insulin administration (0.37 g/min).

Studies with ^{14}C-Labeling

Because of ethical reasons only three studies on the oxidation of exogenous CHO using ^{14}C-labeling have been reported [3–5]. Results reported by Benadé et al. [3] indicate that the oxidation rate of 100 g of sucrose given in the second half of a 6-hour exercise period at 1.6 liters O_2/min, was 0.24 g/min, which fell almost exactly on the regression line computed in figure 1. On the other hand, results reported by Costill et al. [4] and by Van Handel et al. [5] correspond to an oxidation rate as low as 0.05 g/min for $\dot{V}O_2$ ranging from 2 to 2.8 liters/min. No satisfactory explanation has been given for these very low results. In view of the results reported by Benadé et al. [3] using ^{14}C-labeling, and of the consistent results reported using ^{13}C-labeling by various independent authors, a technical problem is suspected in the studies by

Costill et al. [4] and Van Handel et al. [5], which have been conducted in the same laboratory.

Oxidation of Other Exogenous Substrates

It is worth mentioning in this review that Décombaz et al. [31] and Satabin et al. [44] have compared the oxidation rates of medium and long chain triacylglycerols (MCT and LCT) to the oxidation rate of exogenous glucose or maltodextrins under similar experimental conditions. The reported results confirmed the hypothesis underlying both studies, that exogenous fat would be a poor substrate for oxidation during a period of exercise. In fact, the oxidation rate of MCT was between 22 and 33% of that of exogenous CHO, while the oxidation rate of LCT was only 4% of that of exogenous CHO.

Effects of Changes in ^{13}C Background

Methods

As mentioned above, studies conducted before 1990, including our own studies, have failed to take into account changes in the isotopic composition of CO_2 arising from oxidation of endogenous substrates (Rendo) due to both exercise and oxidation of the exogenous substrate. In these studies it was assumed that Rendo in the experimental situation was similar to the isotopic composition of expired CO_2 observed at rest or during exercise without CHO ingestion, i.e. it was assumed that the respective isotopic compositions of the various endogenous substrate stores and pools were similar and/or that the respective contributions of these stores and pools were similar at rest and at exercise, with and without exogenous CHO ingestion. Under this assumption it has been shown by Mosora et al. [46] that the amount of exogenous substrate oxidized (m) is:

$$m = \dot{V}CO_2 \frac{Rexp - Rendo}{Rexo - Rendo} \frac{1}{k} \qquad (1)$$

(see ref. 29 for the derivation of this equation) where $\dot{V}CO_2$ is the production of CO_2; Rexp is the isotopic composition of expired CO_2; m is the amount of the exogenous substrate oxidized; k is the volume of CO_2 provided by the oxidation of one gram of the exogenous substrate; Rexo is the isotopic composition of the exogenous substrate; and Rendo is the isotopic composition of the endogenous substrates.

This equation has been used under this form or other forms, e.g. Ravussin et al. [40], taking Rexp at rest before exercise or in response to exercise without CHO ingestion as the reference value for the background enrichment of expired CO_2, and thus as the value for Rendo. However, as discussed above, due to isotopic fractionation along the metabolic pathways, and, consequently, to differences in isotopic composition of endogenous substrate stores/pools, background enrichment of expired CO_2 is modified when the mixture of endogenous substrates oxidized is modified as, for example, from rest to exercise, and, during exercise, when an exogenous substrate is ingested and oxidized. Accordingly, the value of Rendo in the experimental situation is 'unique' and has to be measured.

We have recently shown [29] that m and Rendo can be computed simultaneously from two observations performed under the same experimental conditions but with ingestion of the exogenous substrate at two different levels of enrichment (Rexo$_A$ and Rexo$_B$). In these two situations, the amount of labeled CO_2 produced are:

$$\dot{V}CO_{2A} = \dot{V}CO_2\,Rexp_A = (\dot{V}CO_2\,endo\,Rendo) + m\,k\,Rexo_A \qquad (2)$$

$$\dot{V}CO_{2B} = \dot{V}CO_2\,Rexp_B = (\dot{V}CO_2\,endo\,Rendo) + m\,k\,Rexo_B \qquad (3)$$

Since it is assumed that $\dot{V}CO_2$, $\dot{V}CO_2$ endo, Rendo and m are similar in the two experiments, equations 2 and 3 constitute a system of two equations with two unknowns (m and Rendo). This system can be solved in order to compute m and Rendo [see ref. 29 for details of computation]:

$$m = \dot{V}CO_2\,\frac{Rexp_A - Rexp_B}{Rexo_A - Rexo_B}\,\frac{1}{k} \qquad (4)$$

$$Rendo = \frac{m\,k\,Rexo_A - \dot{V}CO_2\,Rexp_A}{\dot{V}CO_2 - (m\,k)} \qquad (5)$$

Validation of the Method

This method has been validated on 4 subjects working for 90 min on a cycle ergometer (67% $\dot{V}O_2$ max; $\dot{V}O_2 = 2.8 \pm 0.3$ liters/min). After 30 min of exercise the subjects ingested 30 g of glucose in a single bolus in 300 ml of water. Indirect calorimetry and mass spectrometer analysis of expired CO_2 were performed according to procedures which have been described elsewhere [27]. The exercise was performed three times for each subject. In the first and second experiment small 'natural' enrichments of exogenous glucose were used (Rexo$_A$ = 1.11033 and Rexo$_B$ = 1.12628% or -11.9 and $+2.3$‰ δ ^{13}C/PDB) and the values of m and Rendo were computed using equations 2 and 3. In the third experiment a very high artificial enrichment of exogenous glucose was used ($Rexo_C$ = 1.47595% or +313.5‰ δ ^{13}C/PDB), and the value of m was computed using equation 1, taking Rexp observed after 30 min of exercise as Rendo. Since the expected variations in Rendo are negligible compared to $Rexo_C$, the error made can be considered as small. Results from experiments A and B indicate that over the 60 min of exercise following glucose ingestion Rendo was 1.10008%, compared to 1.09685% at rest and 1.10003% in response to exercise without glucose ingestion, and that 12.9 g of exogenous glucose were oxidized (43% of the dose ingested; average oxidation rate: 0.22 g/min). When the value of m is computed using equation 1, neglecting changes in Rendo, the amount of exogenous glucose oxidized ranges from 16.3 to 57.7 g (54–192% of the dose ingested) depending on the value taken as Rendo (Rexp at rest or in response to exercise without glucose ingestion) and on the enrichment of exogenous glucose used.

Results from experiment C, where m can be estimated while neglecting changes in Rendo due to glucose ingestion, confirm that the actual amount of exogenous glucose oxidized was about 11.7 g.

This pilot experiment indicates that the novel computation procedure is valid and that when natural enrichment of exogenous glucose is used, changes in Rendo cannot be neglected. Failure to take into account these changes may lead to overestimations which can be large, with possibilities of finding more than 100% ^{13}C recovery.

Experimental Results

Using this new computation procedure which takes into account actual changes in background enrichment of expired CO_2 due to exogenous substrate ingestion and exercise, we have computed the oxidation rate of 60 g of exogenous glucose (in 1,000 ml of water) over a 2-hour period of exercise on a bicycle ergometer (67% $\dot{V}O_2$ max) performed by 7 subjects. The glucose solution was ingested in small doses given every 15 min from t = 15 min to t = 105 min. The experiment was repeated twice for each subject ($Rexo_A$ = 1.11482%, $Rexo_B$ = 1.133303%). Oxidation rate of exogenous glucose was related to the energy expenditure (fig. 1). However, exogenous glucose appears to contribute only 6–9% of the energy expenditure compared to 16–20% computed from previous studies presented in figure 1. This observation confirms the notion that when changes in background enrichment of expired CO_2 due to ingestion of exogenous CHO during exercise are not taken into account, the contribution of exogenous CHO to the energy requirement is overestimated. We believe that results published on this question since 1970 should be considered with caution, and that this question should be reinvestigated thoroughly using procedures which adequately take into account changes in Rendo, i.e. the novel procedure we have developed and validated, or the commonly used procedure but using large (and expensive) enrichments of exogenous CHO.

References

1 Coyle EF, Coggan AR: Effectiveness of carbohydrate feeding in delaying fatigue during prolonged exercise. Sports Med 1984;1:446–458.
2 Lamb DR, Brodowicz GR: Optimal use of fluids of varying formulations to minimize exercise-induced disturbances in homeostasis. Sports Med 1986;3: 247–274.
3 Benadé AJS, Wyndham CH, Jansen CR, Rogers GG, de Bruin JP: Plasma insulin and carbohydrate metabolism after sucrose ingestion during rest and prolonged aerobic exercise. Pflügers Arch 1973;342:194–206.
4 Costill DL, Bennett A, Branam G, Eddy D: Glucose ingestion at rest and during prolonged exercise. J Appl Physiol 1973;34:764–769.
5 Van Handel PJ, Fink WJ, Branam G, Costill DL: Fate of ^{14}C glucose ingested during prolonged exercise. Int J Sports Med 1980;1:127–131.
6 Irving CS, Wong WW, Schulman RJ, O'Brian Smith E, Klein PD: (^{13}C) bicarbonate kinetics in humans: intra- vs interindividual variations. Am J Physiol 1983;245: R190–R202.
7 Hoerr RA, Yu Y-M, Wagner D, Burke JF, Young VR: Recovery of ^{13}C in breath from NaH^{13}CO$_2$ infused by the gut and the vein: Effect of feeding. Am J Physiol 1989;257:E426–E438.
8 Robert JJ, Koziet J, Chauvet D, Darmaun D, Desjeux JF, Young VR: Use of ^{13}C labeled glucose for estimating glucose oxidation: Some design considerations. J Appl Physiol 1987;63:1725–1732.
9 Wolfe RR, Wolfe MH, Nadel ER, Shaw JHF: Isotopic determination of amino acid-urea interactions in exercise in humans. J Appl Physiol 1984;56:221–229.
10 Lefèbvre JP, Pirnay F, Pallikarakis N, Krzentowski G, Jandrain B, Mosora F, Lacroix M, Luyckx AS: Metabolic availability of carbohydrates ingested during, before, or after muscular exercise. Diabetes Metab Rev 1986;1:483–500.
11 Bricout J, Fontes JC: Analytical differentiation – cane and beet sugar. Sugar J, March 1977:31–32.
12 DeNiro MJ, Epstein S: Mechanism of carbon isotope fractionation associated with lipid synthesis. Science 1977;197:261–263.
13 Troughton JH: Delta ^{13}C as an indicator of carboxylation reactions; in Gibbs M (ed): Photosynthesis II. New York, Springer, 1979, pp 140–149.
14 Urey HC: The thermodynamic properties of isotopic substances. J Chem Soc 1947: 562–581.
15 Bender MM: Variations in the ^{13}C/^{12}C ratios of plants in relation to the pathway of photosynthetic carbon dioxide fixation. Phytochemistry 1971;10:1239–1244.

16   Ray TB, Black CC: The C_4 and crassulacean acid metabolism pathways; in Gibbs M, Latzko E (eds): Photosynthesis II. New York, Springer, 1979, pp 77–101.

17   Whelan T, Sackett WM, Benedict CR: Carbon isotope discrimination in a plant possessing the C_4 dicarboxylic acid pathway. Biochem Biophys Res Commun 1970;41:1205–1210.

18   DeNiro MJ, Epstein S: Influence of diet on the distribution of carbon isotopes in animals. Geochim Cosmochim Acta 1978;42:495–506.

19   Schoeller DA, Schneider JF, Solomons NW, Watkins JB, Klein PD: Clinical diagnosis with the stable isotope ^{13}C in CO_2 breath tests: methodology and fundamental considerations. J Lab Clin Med 1977;90:412–421.

20   Jacobson BS, Smith BN, Epstein S, Laties GG: The prevalence of carbon-13 in respiratory carbon dioxide as an indicator of the type of endogenous substrate. J Gen Physiol 1970;55:1–17.

21   Mosora F, Lacroix M, Duchesne J: Variations isotopiques $^{13}C/^{12}C$ du CO_2 respiratoire chez le rat, sous l'action d'hormones. C R Acad Sci 1971;273:1752–1753.

22   Duchesne J, Mosora F, Lacroix M, Lefèbvre JP, Luyckx A, Lopez-Habib G: Une application clinique d'une nouvelle méthode biophysique basée sur l'analyse isotopique du CO_2 exhalé par l'homme. C R Acad Sci 1973;277:2261–2264.

23   Lacroix M, Mosora F, Pontus M: Glucose naturally labeled with carbon-13: Use for metabolic studies in man. Science 1973;181:445–446.

24   Massicotte D, Hillaire-Marcel C, Ledoux M, Péronnet F: The natural isotope tracing with ^{13}C: a non invasive method for studying the metabolism during exercise. Can J Appl Sport Sci 1984;9:164.

25   Wolfe RR, Shaw JHF, Nadel ER, Wolfe MH: Effect of substrate intake and physiological state on background $^{13}CO_2$ enrichment. J Appl Physiol 1984;56:230–234.

26   Massicotte D, Péronnet F, Allah C, Hillaire-Marcel C, Ledoux M, Brisson G: Metabolic response to (^{13}C) glucose and (^{13}C) fructose ingestion during exercise. J Appl Physiol 1986;61:1180–1184.

27   Massicotte D, Péronnet F, Brisson G, Bakkouch K, Hillaire-Marcel C: Oxidation of a glucose polymer during exercise: Comparison with glucose and fructose. J Appl Physiol 1989;66:179–183.

28   Massicotte D, Péronnet F, Brisson G, Boivin L, Hillaire-Marcel C: Oxidation of exogenous carbohydrate during prolonged exercise in fed and fasted conditions. Int J Sports Med 1990;11:253–258.

29   Péronnet F, Massicotte D, Brisson GR, Hillaire-Marcel C: Use of ^{13}C-substrates for metabolic studies in exercise: Methodological considerations. J Appl Physiol 1990; 69:1047–1052.

30   Barstow TJ, Cooper DM, Epstein S, Wasserman K: Changes in breath $^{13}CO_2/^{12}CO_2$ consequent to exercise and hypoxia. J Appl Physiol 1989;66:936–942.

31   Décombaz J, Arnaud MJ, Milon H, Moesch H, Philippossian G, Thelin AL, Howald H: Energy metabolism of medium-chain triglycerides versus carbohydrates during exercise. Eur J Appl Physiol 1983;52:9–14.

32   Krzentowski G, Jandrain B, Pirnay F, Mosora F, Lacroix M, Luyckx AS, Lefebvre PJ: Availability of glucose given orally during exercise. J Appl Physiol 1984;56:315–320.

33   Décombaz J, Sartori D, Arnaud MJ, Thelin AL, Schurch P, Howald H: Oxidation and metabolic effects of fructose or glucose ingested before exercise. Int J Sports Med 1985;6:282–286.

34 Gérard J, Jandrain B, Pirnay F, Pallikarakis N, Krzentowski G, Lacroix M, Mosora F, Luyckx AS, Lefèbvre PJ: Utilization of oral sucrose load during exercise in humans: Effect of the α-glucosidase inhibitor acarbose. Diabetes 1986;35:1294–1301.

35 Guézennec CY, Satabin P, Duforez F, Merino D, Péronnet F, Koziet J: Oxidation of corn starch, glucose, and fructose ingested before exercise. Med Sci Sports Exerc 1989;21:45–50.

36 Jandrain B, Krzentowski G, Pirnay F, Mosora F, Lacroix M, Luyckx AS, Lefèbvre PJ: Metabolic availability of glucose ingested 3 h before prolonged exercise in humans. J Appl Physiol 1984;56:1314–1319.

37 Jandrain BJ, Pirnay F, Lacroix M, Mosora F, Scheen AJ, Lefèbvre PJ: Effect of osmolality on availability of glucose ingested during prolonged exercise in humans. J Appl Physiol 1989;67:76–82.

38 Krzentowski G, Pirnay F, Luyckx AS, Lacroix M, Mosora F, Lefèbvre P: Effect of physical training on utilization of glucose load given orally during exercise. Am J Physiol 1984;246:E412–E417.

39 Krzentowski G, Pirnay F, Pallikarakis N, Luyckx AS, Lacroix M, Mosora F, Lefèbvre PJ: Glucose utilization during exercise in normal and diabetic subjects: the role of insulin. Diabetes 1981;30:983–989.

40 Pallikarakis N, Jandrain B, Pirnay F, Mosora M, Lacroix M, Luyckx A, Lefèbvre P: Remarkable metabolic availability or oral glucose during long-duration exercise in humans. J Appl Physiol 1986;60:1035–1042.

41 Pirnay F, Crielaard JM, Pallikarakis N, Lacroix M, Mosora F, Krzentowski G, Luyckx AS, Lefèbvre PJ: Fate of exogenous glucose during exercise of different intensities in human. J Appl Physiol 1982;53:1620–1624.

42 Pirnay F, Lacroix M, Mosora F, Luyckx A, Lefèbvre PJ: Glucose oxidation during prolonged exercise evaluated with naturally labeled (^{13}C) glucose. J Appl Physiol 1977;43:258–261.

43 Ravussin E, Pahud P, Dorner A, Arnaud MJ, Jéquier E: Substrate utilization during prolonged exercise preceded by ingestion of ^{13}C-glucose in glycogen depleted and control subjects. Pflügers Arch 1979;382:197–202.

44 Satabin P, Portero P, Defer G, Bricout J, Guézennec CY: Metabolic and hormonal responses to lipid and carbohydrate diets during exercise in man. Med Sci Sports Exerc 1987;19:218–223.

45 Wagenmakers AJM, Brouns F, Saris WHM, Halliday D: Maximal oxidation of oral carbohydrates during exercise (abstract). Med Sci Sports Exercise 1990; S120.

46 Mosora F, Lefèbvre P, Pirnay F, Lacroix M, Luyckx A, Duchesne J: Quantitative evaluation of the oxidation of an exogenous glucose load using naturally labeled ^{13}C-glucose. Metabolism 1976;25:1575–1582.

Dr. François Péronnet, Département d'éducation physique, Université de Montréal, CP 6128-A, Montréal, P.Q. H3C3J7 (Canada)

Marconnet P, Komi PV, Saltin B, Sejersted OM (eds): Muscle Fatigue Mechanisms in Exercise and Training. Med Sport Sci. Basel, Karger, 1992, vol 34, pp 207–217

Heat Stress Causes Fatigue!

Exercise Performance during Acute and Repeated Exposures to Hot, Dry Environments

Bodil Nielsen

August Krogh Institute, University of Copenhagen, Denmark

Introduction

In hot compared to cool conditions the endurance for prolonged exercise is reduced. In other words, heat stress leads to an earlier occurrence of fatigue.

On the other hand, repeated exposure to hot environments leads to acclimatization, the ability to withstand the stress is augmented, the fatigue is delayed. The causes for this are to be discussed here, and some of the possibilities are mentioned below.

The competition for the cardiac output with the increased demand for skin circulation in hot environments, together with the metabolic requirements of blood flow to the exercising muscles is met by a reduction in renal, splanchnic and hepatic blood flow [1–4]. At low-to-moderate exercise intensities cardiac output may increase in hot environments [5–7]. In spite of this, the result might be a *reduction in muscle blood flow*, and an *altered muscle metabolism*, which could contribute to the early fatigue.

Further, the fluid loss due to sweating might *reduce plasma volume* and increase the circulatory strain, – and the electrolyte loss with the sweat may alter the *electrolyte balance* around the cells (muscle and nerve) and affect their function, – the *high muscle temperature* could interfere with muscle metabolism – and, finally, the excessively *high core temperature* might in itself cause the fatigue and inability to continue exercise during heat stress.

Circulatory Responses: Previous Studies

Indications of reduced muscle blood flow during exercise and heat exposure was found by Donald et al. [8] and Williams et al. [9] from measurements of *femoral venous O_2 content*, which was reduced in heat stress.

Further, McDougal et al. [10] observed in 6 subjects running at 65–70% of max $\dot{V}_{O_2}$ on a treadmill in water-perfused suits perfused at 18 °C, 23 °C or at rectal temperature, that the *lactate concentration* in venous blood increased with increasing heat stress. This too could be an indication of a reduction in blood flow to the exercising muscle, and/or a decrease in rate of lactate removal. The subjects became exhausted earlier the higher the heat stress. It occurred at about 39.5 °C rectal temperature in all 3 conditions.

Similarly, subjects bicycling, at 70–85% of $\dot{V}_{O_2}$ max at 9 °C and 41 °C air temperature, had an increased venous lactate concentration, and an enhanced glycogen breakdown in biopsies from the active muscles in the hot condition [11].

Measurements of muscle blood flow in animal experiments were done by Bell et al. [12]. They studied sheep exercising on a treadmill in thermoneutral and hot environments. With a microsphere technique, they demonstrated a reduction in blood flow in the leg muscles in hot compared to thermoneutral conditions.

However, the first direct measurements of human leg blood flow during heat stress in one-legged knee extension, seated and upright two-legged bicycle exercise up to 40–50% of $\dot{V}_{O_2}$ max showed no significant changes in leg blood flow, measured with thermodilution in the femoral vein. The heat stress was imposed by a water-perfused suit [13].

Metabolic Changes with Heat Stress

Also with heat exposure, metabolic change could take place in exercising muscles. The excessive rise in muscle temperature could contribute to the increased lactate formation and the higher glycogen utilization [11]. In dogs, Kozłowski et al. [14] found that when the hyperthermia occurring in prolonged exercise was reduced by cooled vests, the endurance time was increased, and the metabolic breakdown of glycogen, lactate accumulation, and ATP depletion was delayed.

We have studied these mentioned aspects of fatigue due to heat stress, during *acute heat exposure*, [15] – and the changes evoked by *repeated heat exposures*, an 8–12 days acclimation to dry heat, 40–42 °C. The measurements included circulatory parameters, and metabolic and hormonal variables, in an attempt to judge both local conditions in the exercising muscle, and whole body reactions.

Acute Heat Exposure: Procedure and Results

The subjects exercised by walking uphill on a treadmill at 60% of their $\dot{V}_{O_2}$ max. After 30 min in a cool environment (18 °C) they moved to an adjacent room (at 40 °C) and continued to exercise at the same speed and inclination until exhaustion, or 60 min whichever occurred first.

Measurements taken at steady state after 30 min in cool condition and the final measurements at exhaustion are compared in order to find the causes for the inability of the subjects to continue the exercise.

Core temperature (esophageal, T_{es}) which had reached a level at 38.2 °C in the cool condition, increased in the hot environment to a final temperature of 39.3 ± 0.2 °C. Whole body oxygen uptake ($\dot{V}_{O_2}$ had increased, and cardiac output tended to increase. Heart rate (HR) rose from 146 ± 3 to a near maximal value of 183 ± 3.

The blood flow from the leg was not changed, nor was the local (a–v)O_2 difference, and local O_2 uptake in the leg (table 1). These last mentioned are

Table 1. Measurements during exercise in cool and hot conditions

	Cool 25–30 min	Hot, final 75–90 min
T_{es}, °C	38.2 ± 0.1	39.3 ± 0.1*
$\dot{V}_{O_2}$, liters · min^{-1}	2.38 ± 0.12	2.82 ± 0.09*
HR, bpm	146 ± 5	183 ± 3*
Cardiac output, liters · min^{-1}	15.2 ± 0.7	18.4 ± 1.1
Leg blood flow, liters · min^{-1}	6.1 ± 0.8	6.0 ± 0.5
Leg (a–v)O_2 difference, ml · l^{-1}	145.6 ± 5.08	147.9 ± 3.73
Leg $\dot{V}_{O_2}$, liters · min^{-1}	0.89 ± 0.13	0.91 ± 0.11
Lactate (femoral vein), mM	1.3 ± 0.1	2.2 ± 0.2
pH (femoral vein)	7.29 ± 0.01	7.33 ± 0.01

Values are means and ± SE for 7 subjects, except for cardiac output where n = 5. From Nielsen et al. [15].
* Significantly different from 30-min value.

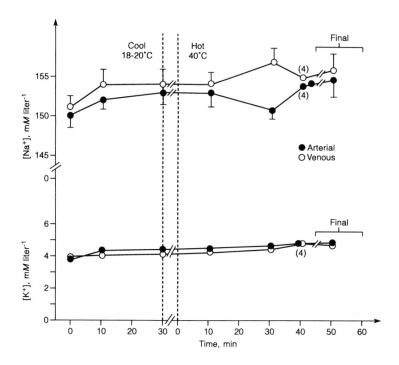

Fig. 1. Arterial and femoral vein Na⁺ and K⁺ concentrations during 30 min walking in cool and up to 60 min in hot environment [data from ref. 15]. Seven subjects, mean values and ±SE are presented.

indications that the *blood flow* and oxygen supply to the exercising leg is *not reduced*, and cannot be the reason for the exhaustion.

This is in contrast to the experiments mentioned earlier [3, 8, 9, 11] where, from measurements of increased lactate production and glycogen breakdown during exercise in hot conditions, it was deduced that flow reduction in working muscles takes place.

In our experiments plasma volume decreased at the onset of exercise. Thereafter, it increased gradually, and had returned to the resting value after 30 min in the cool condition. At the end of exercise in the hot environment it was slightly above the initial level. Therefore, reduction in plasma volume is not a cause for the fatigue. Nor were the arterial or venous electrolyte concentrations extreme: Na⁺ (153 ± 2 at 30 min, 155 ± 1 mM final) and K⁺ (from 3.9 ± 0.07 to final 5 ± 0.11 mM) in the femoral vein (fig. 1).

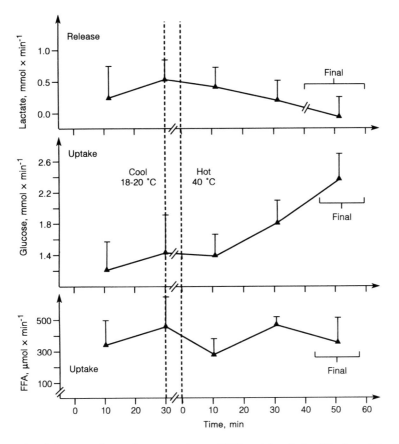

Fig. 2. Lactate release, glucose uptake, and FFA uptake, during 30 min walking in a cool environment, continued for up to 60 min in a hot environment. Seven subjects, mean values and ±SE are presented [from ref. 15].

The substrate utilization in the exercising leg is illustrated in figure 2. Glucose and free fatty acid uptake was not reduced by the exercise in hot environments, and lactate release remained low, close to zero in the final sample in our experiments.

Muscle biopsies were taken at rest, after 30 min in cool condition and at exhaustion. The muscle glycogen stores were not depleted when the subject had to stop, and the muscle glycogen utilization rate: 0.60 mmol/min · kg wet weight in the cool condition, and 0.40 mmol/min · kg wet weight during heat exposure, was not increased during heat stress (in contrast to the mentioned findings [11, 14]).

We therefore conclude that neither flow limitation to working muscles nor altered metabolism or substrate depletion is the cause for the fatigue in the present circumstances. We believe that high core temperature in itself causes fatigue [15, 16].

Acclimatization to Hot Environments

Repeated exposures to hot environments lead to *acclimatization*: the ability to withstand the stress is augmented. The question is then: What are the changes that occur with acclimatization, and in which way do these changes delay the fatigue? What is the ultimate cause for exhaustion after acclimation? *Acclimation* is the term used for the physiological changes which occur in response to experimentally induced changes in one particular climatic factor such as ambient temperature, as opposed to *acclimatization* which is used to describe the reactions to the natural climate. The process has been extensively studied with respect to changes in central circulation, temperature regulation and in body fluid spaces. It is, however, still under debate as to which changes are primary or secondary in the adaptive processes, and also the results of various studies are different.

In earlier studies it was found that the *central circulation* is affected. At the first heat exposure stroke volume is decreased and heart rate (HR) increased compared to control values. Through subsequent heat exposures the stroke volume tends to increase while HR is gradually lowered. The resulting cardiac output may decrease [17], increase [18] or remain unchanged in relation to control values [19, 20]. The variable results are probably due to different acclimation procedures, exercise conditions and intensities, and different effects of hot wet and hot dry conditions.

The *skin circulation* estimated by conductance of peripheral tissues is gradually reduced with repeated heat exposures [20, 21] or found to be unchanged [22]. Measured with venous occlusion plethysmography forearm blood flow (FBF) remained unchanged [23]. Roberts et al. [24] reported that acclimation led to a reduced threshold for vasodilation, which occurred at a lower core temperature during exercise after acclimation. Thus, at a given core temperature FBF was slightly higher. But since core temperature is lower after acclimation, the results may indicate a reduction in total skin circulation after acclimation.

The splanchic and renal blood flow which is reduced during acute heat exposure is assumed to increase after acclimation, supposing that skin blood flow is decreased and cardiac output is unaffected after acclimation [25, 26].

According to King et al. [27], the blood flow to exercising muscle may have increased after acclimation since the glycogen utilization was reduced. This they ascribe also to an increased hepatic glucose release, and therefore an enhanced exogenous substrate supply to the exercising muscle after acclimation, because lactate formation and blood lactate concentrations were not different before and after acclimation.

Our preliminary results with direct measurements of leg blood flow support this assumption. We find a tendency for an increased muscle blood flow after acclimation.

Sweat rate is found to increase with acclimation. This is confirmed in our ongoing study. The change in sweat rate can be described as an increased sensitivity and a lower threshold for onset of sweating with acclimation [22, 25, 27, 28]. This produces a lowering of skin temperature, reduces the need of skin blood flow for heat transfer, and results in a lowering of core temperature.

The underlying mechanisms for the adaptive changes are still under debate [29].

Acclimation: Dry Heat, Procedure and Results

In ongoing experiments we are investigating the *cardiac output, and its distribution* before and after acclimation. So far, 6 very-well-trained subjects have been examined. An initial experiment was performed in cool conditions (20 °C), in which the work intensity was adjusted to a HR of 125–130 bpm so that the subjects could exercise for 90 min. Thereafter, they exercised daily in a hot, dry condition (40–42 °C ambient air and wall temperature, about 10% Rh) at the same predetermined work load until exhaustion or 90 min. In the present context we have analysed the results for an explanation of the delayed onset of fatigue with acclimation. We therefore compare values from the first to last heat exposure at the point of exhaustion, when the subjects were unable to continue the exercise.

The cardiac output had increased in all subjects, on an average by 2.2 liters/min, due to an increased stroke volume.

Leg blood flow also tended to increase, indicating an increased muscle circulation. For the two legs together the increase was about 0.8 liters/min. Further, the skin blood flow measured by plethysmography had increased by 6.6%.

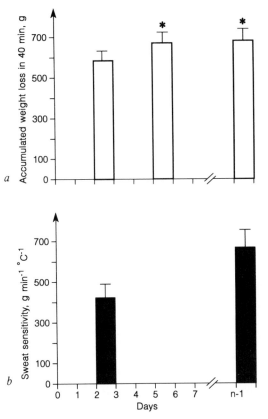

Fig. 3. a Weight loss in 40 min on the 2nd, 5th and final but 1 day during exercise in 40–42 °C dry heat. Mean and SE, 6 subjects. *b* Sweat sensitivity, the weight loss per °C increase in esophageal temperature on the second and final but 1 day of exercise in 40–42 °C dry heat. Mean and SE, 6 subjects.

Sweat rate was also increased. The rate of weight loss which is a measure of the evaporation heat loss, and the sensitivity of the sweat mechanism: g sweat per min per degree rise in core temperature is augmented (fig. 3).

Neither metabolite concentrations, utilization of these, nor the water and electrolyte-regulating hormone concentrations had changed due to the acclimation period, and thus could not have contributed to the better performance after acclimation, or to the delayed exhaustion.

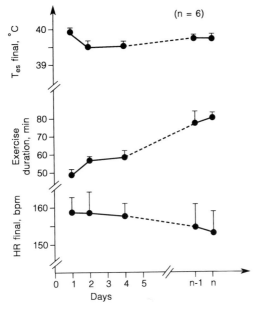

Fig. 4. Endurance time, esophageal temperature (T_{es}) and HR at exhaustion in 6 subjects exercising for 9–12 consecutive days in 40–42 °C dry heat. Mean and SE, 6 subjects.

The time at which exhaustion occurred corresponded to the time when HR reached about 169 bpm and T_{es} about 40 °C. This took 49 ± 2 min (average of 6) on the first day and 80 ± 3 min (average of 6) on the last day (fig. 4).

Conclusion

The cardiovascular system adapts to repeated heat and exercise stress by an increased cardiac output and stroke volume. The increased skin blood flow and sweating sensitivity together lead to a slower rise in core temperature.

We found no reduction in muscle blood flow during exercise till exhaustion. It was *the rise in core temperature to about 40 °C which elicited, or coincided with exhaustion and inability to continue to exercise* both during acute heat exposure, and after acclimation to heat.

High core temperature is therefore the ultimate cause for the fatigue due to heat stress. It may reduce the function of motor centers, its ability to recruit motor units for the required work, or perhaps affect 'motivation' for motor activity.

References

1   Radigan LR, Robinson S: Effects of environmental heat stress and exercise on renal blood flow and filtration rate. J Appl Physiol 1949;2:185–191.
2   Rowell LB, Blackmon JR, Martin RH, Wazarella JA, Bruce RA: Hepatic clearance of indocyanine green in man under thermal and exercise stresses. J Appl Physiol 1965; 20:384–394.
3   Rowell LB, Brengelmann GL, Blackmon JR, Twiss RD, Kusumi F: Splanchnic blood flow and metabolism in heat-stressed man. J Appl Physiol 1968;24:475–484.
4   Rowell LB: Human Circulation Regulation during Physical Stress. New York, Oxford University Press, 1986.
5   Klausen K, Dill DB, Philips EE Jr, McGregor D: Metabolic reactions to work in the desert. J Appl Physiol 1967;22:292–296.
6   Nadel ER, Cafarelli E, Roberts MF, Wenger CB: Circulatory regulation during exercise in different ambient temperatures. J Appl Physiol 1979;46:430–437.
7   Rowell LB, Murray JA, Brengelmann GL, Kraning KK: Human cardiovascular adjustments to rapid changes in skin temperature during exercise. Circulation Res 1969;24:711–724.
8   Donald KW, Wormald PN, Taylor SH, Bishop JM: Changes in the oxygen content of femoral venous blood and leg blood flow during leg exercise in relation to cardiac output response. Clin Sci 1957;16:567–591.
9   Williams CG, Bredell GAG, Wyndham CH, Strydom NB, Morrison JF, Peter J, Fleming PW: Circulatory and metabolic reactions to work in heat. J Appl Physiol 1962;17:625–638.
10 McDougall JD, Reddan WG, Layton CR, Dempsey JA: Effects of metabolic hyperthermia on performance during heavy prolonged exercise. J Appl Physiol 1974;36: 538–544.
11 Fink WJ, Costill DL, Van Handel PJ: Leg muscle metabolism during exercise in the heat and cold. Eur J Appl Physiol 1975;34:183–190.
12 Bell AW, Hales JRS, King RB, Fawcett AA: Influence of heat stress on exercise-induced changes in regional blood flow in sheep. J Appl Physiol 1983;55:1916–1923.
13 Savard GK, Nielsen B, Laszczynska I, Larsen BE, Saltin B: Muscle blood flow is not reduced in humans during moderate exercise and heat stress. J Appl Physiol 1988;64: 649–657.
14 Kozłowski S, Brzezinska Z, Kruk B, Kaciuba-Uscilko H, Greenleaf JE, Nazar K: Exercise hyperthermia as a factor limiting physical performance: temperature effect on muscle metabolism. J Appl Physiol 1985:59:766–773.
15 Nielsen B, Savard G, Richter EA, Hargreaves M, Saltin B: Muscle blood flow and muscle metabolism during exercise and heat stress. J Appl Physiol 1990:69:1040–1046.

16 Brück K, Olschewski H: Body temperature related factors diminishing the drive to exercise. Can J Physiol Pharmacol 1987;65:1274–1280.

17 Wyndham CH: Effect of acclimatization on circulatory responses to high environmental temperature. J Appl Physiol 1951;4:383–395.

18 Wyndham CH, Rogers GG, Senay LC, Mitchell D: Acclimatization in a hot, humid environment: Cardiovascular adjustments. J Appl Physiol 1976;40:779–785.

19 Rowell LB, Kraning KK, Kennedy JW, Evans TO: Central circulatory responses to work in dry heat before and after acclimatization. J Appl Physiol 1967;22:509–518.

20 Wyndham CH, Benade AJA, Williams CG, Strydom NB, Goldin A, Heyns AJA: Changes in central circulation and body fluid spaces during acclimatization to heat. J Appl Physiol 1968;25:586–593.

21 Eichna LW, Park CR, Nelson N, Horvath SM, Palmes CD: Thermal regulation during acclimatization in a hot dry (desert-type) environment. Am J Physiol 1950; 163:585–597.

22 Mitchell D, Senay LC, Wyndham CH, van Rensburg AJ, Rogers GG, Strydom NB: Acclimatization in a hot humid environment: energy exchange, body temperature and sweating. J Appl Physiol 1976;40:768–778.

23 Whitney RJ: Circulatory changes in the forearm and hand of man with repeated exposure to heat. J Physiol London 1954;125:1–24.

24 Roberts MF, Wenger CB, Stolwijk JAJ, Nadel ER: Skin blood flow and sweating changes following exercise training and heat acclimation. J Appl Physiol 1977;43: 133–137.

25 Rowell LB: Human cardiovascular adjustments to exercise and thermal stress. Physiol Rev 174;54:75–159.

26 Rowell LB: Cardiovascular adjustments to thermal stress. Handbook of Physiology. The Cardiovascular System II. USA, 1983.

27 King DS, Costill DL, Fink WJ, Hargreaves M, Fielding RA: Muscle metabolism during exercise in the heat in unacclimatized and acclimatized humans. J Appl Physiol 1985;55:1107–1110.

28 Candas V: Adaptation to extreme environments. Thermophysiological changes in man during humid heat acclimation; in Dejours P (ed): Comparative Physiology of Environmental Adaptations, vol 2. Karger, Basel 1987, pp 76–93.

29 Senay LC: An inquiry into the role of cardiac filling pressure in acclimatization to heat. Yale J Biol 1986;29:247–256.

Bodil Nielsen, PhD, August Krogh Institute, University of Copenhagen,
9 Blegdamsvej, DK–2100 Copenhagen Ø (Denmark)

Marconnet P, Komi PV, Saltin B, Sejersted OM (eds): Muscle Fatigue Mechanisms in
Exercise and Training. Med Sport Sci. Basel, Karger, 1992, vol 34, pp 218–238

Circadian Rhythms in Muscular Activity

Thomas Reilly

Centre for Sport and Exercise Sciences, Liverpool Polytechnic, Liverpool, UK

Introduction

Life depends on sequences of physiological events that are reproducible
(with greater or lesser precision) at regular intervals and that are evident even
when environmental conditions are constant. The sequences represent biological rhythms which are mainly a product of internal clocks or oscillators
that bestow cyclical activity on physiological processes. The study of rhythms
and biological clocks is known as chronobiology.

Biological rhythms are characterized by their length: for example, a
rhythm that fits the solar day closely is known as circadian, or about a day. Its
period is 24 h approximately. The other characteristics of rhythm are its
amplitude or mean to peak variation, its mesor or mean value and its
acrophase or time that the peak occurs (fig. 1). The idealized rhythm is
represented by a sine wave and hence cosinor analysis is the main statistical
method of identifying circadian rhythms [Reilly, 1990a].

Some rhythms may originate from endogenous sources and are maintained when environmental conditions are constant: others may be attributable largely to exogenous factors, that is, the rhythm is a direct result of
changes in the environment. The most prominent environmental changes are
those arising from the regular spin of the earth around its central axis,
thereby alternating day and night. The solar day dictates habits of sleep, rest
and activity, work and social activity [De Looy et al., 1988]. Light is an
important factor in determining rhythm characteristics and it can be mani-
pulated to alter rhythms.

The major rhythms of relevance for muscle activity are those of body
temperature (fig. 2) and of central nervous system arousal associated with the
sleep-wake cycle. The location of the oscillator controlling these cycles is
deemed to be in the suprachiasmatic nucleus cells of the hypothalamus while

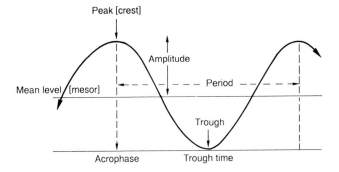

Fig. 1. Terms used to describe a classical circadian rhythm.

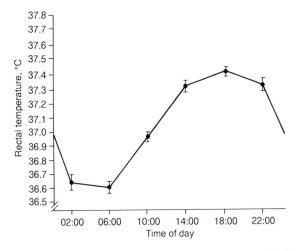

Fig. 2. The circadian variation in rectal temperature [Reilly, 1987].

the pineal gland is also thought to play a role. In addition to these master clocks there are also local timekeepers, conceivably operating within a hierarchy of clocks [Minors and Waterhouse, 1981a]. These include time-keepers within skeletal and cardiac muscle.

This paper will focus on circadian rhythms and muscular activity as they are reflected in exercise performance. First all-out performance, whether competitive or time-trial, is considered. Motor performances which comprise discrete components of exercise are then examined. The physiological responses to exercise are discussed in the context of precursors of fatigue. Thermoregulation and physical factors associated with training are then dealt with. Sub-harmonics within the normal circadian rhythms are also examined. Finally, the disruption of circadian rhythmicity, by means of phase shifts and sleep loss, is treated.

Circadian Variation in Motor Performance

In a study in our laboratory of 17 motor performance measures considered to be important components of sport, 12 showed a typical cosinor curve. The majority of these measures varied with time of day in close agreement with the variation in body temperature. They included grip strength, simple reaction time, vertical jump and speed of a 5-min shuttle run. This suggests that isometric muscle force, dynamic muscle activity, muscle power, neuromotor activity and gross motor performance are all affected by circadian rhythmicity. So also was muscular co-ordination, as reflected in a pursuit rotor test and in an agility run [Stockton et al., 1978].

The circadian rhythm in dynamic activity has been confirmed in other studies using either standing broad jump or vertical jump [Reilly and Down, 1986a]. The variation in such tests attributable to time of day represents about 3% of the mean value. This may be greater than the measurement error in contemporary dynamic dynamometry and would explain why clear circadian rhythms are not always evident in performance of tests such as Wingate Anaerobic Power Test [Reilly and Down, 1986b]. Rhythms have not been apparent in measurements of maximum concentric and maximum eccentric strength of the knee extensors using isokinetic dynamometry [Cabri et al., 1988].

A circadian rhythm in grip strength has been demonstrated by a number of investigators [Colquhoun, 1971; Reilly, 1987]. It has also been shown to persist under conditions where subjects were deprived of sleep for 4 consecutive nights [Reilly and Walsh, 1981]. The circadian variation in grip strength is much greater than the effect of sleep loss on this function [Reilly and Hales, 1988]. Although the rhythm is roughly in phase with that of body temperature, it is not clear whether variations in muscle temperature or in the motivational level of the subjects are mainly responsible for it. According to Åstrand and Rodahl [1986] muscle performance is optimal at a temperature of 38.3 °C, this being generally achieved by competitive athletes as a result of 'warming up'. Experimental work using electrical stimulation of muscle is generally conducted at a constant muscle temperature, a procedure that might swamp an inherent circadian rhythm. A satisfactory definitive study comparing maximal voluntary efforts at a range of times during the solar day with muscle performance under electrical stimulation has not yet been conducted.

Not all rhythmic fluctuations exactly fit the classical pattern. Tapping speed – moving a stylus as rapidly as possible between two plates – showed a slight dip in performance in the early afternoon [Stockton et al., 1978]. This is known as the 'post-lunch dip', although it persists in some tasks even when no lunch is taken. A similar fall in isometric strength of the quadriceps was reported by Wit [1980], measurements being repeated at 1-hour intervals throughout the period of normal wakefulness. Such transient declines in performance during the day may be attributable to an underlying 90-min cycle that prevails during sleep and comes closest to the surface of detection during the day in the early afternoon. In experimental measurements of maximal muscle function, the drop in performance is usually observed when tests are repeated throughout the day, and so it may reflect a fall in arousal or motivational level. The data for tapping speed were obtained on a single subject monitored longitudinally over a 3-month span. This is comparable with that for isometric strength of the knee extensors measured repeatedly every 1 h within a 24-hour period by Wit [1980]. Such measurement sequences may be more directly related to industrial shift work rather than sports performance, the latter being usually concentrated within a relatively short period of the solar day.

Baxter and Reilly [1983] also noted exceptions to the classical circadian rhythm when they examined swimming performances over 100 m and 400 m distances. Times for both all-out swim trials improved steadily throughout the day from 06:00 h on. There was no turning point evident before the final measurements at 22:00 h. The study indicated that evening time was best for sprint swimmers, especially if time trial results are considered important, such as in achieving championship qualifying standards.

The majority of athletic records are set in the late afternoon or evening, close to the time that body temperature is at its peak. All the world records set by 4 British runners at distances of 800–5,000 m in the 10 years between 1978 and 1988 were achieved between 19:00 and 23:00 h. The only track and field records set in the morning since World War II have been in 'explosive' field events – men's shot and women's javelin [Reilly, 1987]. The strong neural component may have been an important factor. Gross muscular strength generally follows the curve of body temperature but powerful actions requiring a suppression of normal neural inhibition may have an earlier acrophase. This may apply also to speedy actions with a cognitive component, as noon is the optimal time of day for certain cognitive functions [Folkard, 1975].

Care should be taken in making inferences from athletic performance records. Firstly, data should be weighted for frequency of organization distributed throughout the day. Secondly, the more favorable temperatures on summer evenings may partly account for the occurrence of best performances at the time that body core temperature is normally at its high point. Nevertheless, athletes do prefer this time of day for record attempts. Their judgements have been vindicated from the first systematic study of diurnal variation in simulated sports competition by Conroy and O'Brien [1974]. Six runners, 3 weight throwers and 3 oarsmen were found to do better in the evening than in the morning. The same research group found that swimmers produced faster times over 100 m at 17:00 h compared with 07:00 h in 3 of 4 strokes studied [Rodahl et al., 1976]. Although there were insufficient measurement points to define the characteristic of a rhythm, the magnitude of change was similar to that observed later by Baxter and Reilly [1983] in their study of swimmers. The steady improvement throughout the day amounted to 3.5% for 100 m and 2.5% for 400 m.

Rhythms have been identified in many components of athletic performance including sensory motor (simple reaction time), psychomotor (hand-eye coordination), sensory perceptual, cognitive, neuromuscular, psychological, cardiovascular and metabolic functions. It is incorrect to speak of a single performance rhythm since different types of task can exhibit different circadian performance rhythms, depending largely on whether they are influenced by the temperature or sleep-wake oscillator [Shephard, 1984; Winget et al., 1985]. Nevertheless, it appears that there is a window around the peak time of body temperature when the state of the body favors maximal performance. Although individuals do vary in the phase and amplitude of their rhythms, such variability may be small [Reilly, 1990b].

Physiological Responses to Exercise: Rhythms at Rest

The physiological responses to exercise, both at submaximal and maximal intensities, may contain an influence of circadian rhythm in functions that determine exercise performance. Since the factors affecting performance differ greatly according to its anaerobic and aerobic components, teasing out

Table 1. Resting rhythm in longitudinal and transverse studies

Variable	Amplitude (% of mean)	Acrophase (decimal clock hours)
Heart rate	6.0–6.1	13.84–15.51
$\dot{V}_E$	7.0–9.7	16.65–17.02
$\dot{V}_{O_2}$, liters	6.4–6.5	17.38–17.43
$\dot{V}_{O_2}$, ml/kg	6.6–6.7	17.20–17.22
Rectal temperature	0.6–0.8	17.74–19.43

the influence of circadian rhythms is not easy. Besides, during exercise the metabolism is elevated many times over the resting level and the amplitude of the circadian rhythm existing at rest is consequently reduced when calculated as a percent of the maximal exercise value. Before rhythms in physiological functions important in evaluating the responses to exercise are considered, their cyclic characteristics are first described for resting conditions.

Circadian rhythms have been reported in endocrine functions including anterior pituitary, adrenal medulla and adrenal cortex hormones. Hormones associated with arousal (e.g. adrenaline and noradrenaline) are elevated during the day compared with night-time values [Minors and Waterhouse, 1981a]. The greater blood volume reported during the night compared to daytime and the reduced vasoconstriction does not seem to be beneficial to muscle function. Glucose tolerance is impaired in the afternoon, although blood glucose levels tend to be higher at that time than in the morning [Zimmet et al., 1974]. Muscle glycogen also shows diurnal variation which cannot be explained by hormonal or metabolic rhythms [Conlee et al., 1976], suggesting the existence of a local timekeeper within skeletal muscle.

Various studies have shown rhythms in renal, circulatory, ventilatory and metabolic functions at rest [Minors and Waterhouse, 1981a]. A rhythm in blood pressure is also observed but usually only when intra-arterial measurements are made. The amplitude of the rhythm in heart rate is 4 beats min^{-1} [Reilly et al., 1984]. Consistently, our observations (table 1) show that the rhythm in heart rate leads the rhythm in oxygen uptake ($\dot{V}_{O_2}$) and ventilation ($\dot{V}E$) which in turn shows a phase lead over core temperature [Reilly and Brooks, 1982, 1990]. The rhythms in motor performance tend to follow the body temperature curve more closely than the other rhythms. About 40% of the rhythm in $\dot{V}_{O_2}$ and 25% of that in $\dot{V}_E$ can be explained by a Q_{10} effect of body temperature. Generally, the amplitude of these rhythms is about 7–11% of the mean value, except for body temperature, the variation of which is best expressed by reference to its physiological range [Reilly, 1987].

Responses to Submaximal Exercise

Our laboratory protocol for examining circadian variation in responses to exercise entails firstly a 10-min rest pre-exercise, then two submaximal steady-rate exercise intensities followed by a graded exercise test to exhaustion. Recovery measurements may be made postexercise. The physiological cost of exercise can be examined by subtracting resting values and rhythm characteristics observable at rest can be compared with those persisting during exercise and recovery. Diet and physical activity as well as environmental conditions have to be controlled during such experiments.

The rhythm in $\dot{V}_{O_2}$ apparent at rest gradually fades as exercise intensity is raised. The rhythm at moderate exercise levels is accounted for by variation in body mass [Reilly and Brooks, 1982]. No circadian variation is evident in $\dot{V}_{CO_2}$ or respiratory exchange rate (RER). This suggests that choice of substrate as fuel for exercising muscle is not determined by time of day, once diet, environmental temperature, prior activity and exercise intensity are controlled.

The rhythm in $\dot{V}_E$ is amplified at light and moderate exercise intensities. This is due to a combination of autonomic and endocrine mechanisms, narrower airways increasing the resistance to breathing at the cooler body temperatures. Even when $\dot{V}_E$ is expressed as ventilation equivalent of oxygen, the rhythm is clearly evident [Reilly, 1982]. This may partly explain the mild dyspnea sometimes associated with exercising in the early morning. The ventilation threshold – which indicates the point where ventilation begins to increase disproportionately to $\dot{V}_{O_2}$ – does not vary with time of day.

The heart rate rhythm noted at rest is still evident at light and moderate exercise intensities. The consistency of its phase and amplitude has been shown for both arm [Cable and Reilly, 1987] and leg [Reilly et al., 1984] exercise. It is not known if the lower heart rate at night-time compromises blood flow to active muscles.

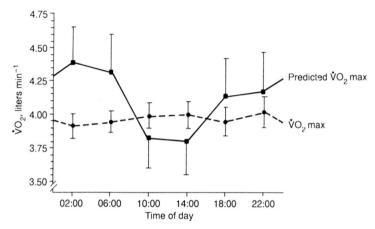

Fig. 3. The contrast between results measured for $\dot{V}_{O_2}$ max throughout the day and values predicted from heart rate response to submaximal exercise, from Reilly [1987].

Maximal Physiological Functions

When maximal values are being attributed to measurements, the question arises as to whether the ceiling of physiological capacity was reached during the exercise test. Consequently, recognized criteria are applied when assessing $\dot{V}_{O_2}$ max. Otherwise data collected during graded exercise to volitional exhaustion may merely reflect the reluctance of subjects to work at $\dot{V}_{O_2}$ max at night-time. For a well-trained individual with a body weight of about 70 kg, the normal amplitude of the $\dot{V}_{O_2}$ rhythm at rest would be less than 0.5% of the mean maximal value. It is difficult to detect that against the background of biological variation and measurement error associated with assessing $\dot{V}_{O_2}$ max. It should not be surprising, therefore, that many carefully conducted studies fail to observe circadian variation in $\dot{V}_{O_2}$ max. We have used both a longitudinal design to eliminate variability between subjects as well as a cross-sectional approach: the coefficient of variation of $\dot{V}_{O_2}$ max was found to be 2.9% [Reilly and Brooks, 1982]. In these studies it was concluded that $\dot{V}_{O_2}$ max is a stable function, independent of time of day. This is in sharp contrast to the value predicted from submaximal heart rate, which shows an error in estimating $\dot{V}_{O_2}$ max that is not acknowledged when the maximal function is predicted from a submaximal test (fig. 3).

The circadian rhythms in $\dot{V}_E$, apparent during light and moderate exercise, similarly disappears under maximal aerobic conditions [Reilly and

Brooks, 1982, 1990]. In arm exercise the highest metabolic measurements generally do not demonstrate a plateau and are referred to as peak rather than maximal values. During performance of arm ergometry the $\dot{V}_{O_2}$ peak, $\dot{V}_E$ peak and highest heart rate have been found to demonstrate a circadian rhythm, the highest values being observed close to the crest time of rectal temperature. The result reflected a rhythm in the total work performed in the incremental arm-exercise test to exhaustion [Cable and Reilly, 1987].

The maximal heart rate does consistently show an effect of time of day. The rhythm is similar in phase to that noted at rest and submaximal exercise but its amplitude is reduced [Reilly and Brooks, 1990]. This is as expected of any biological function as it approaches the limit of its physiological range. The rhythm may reflect a thermogenic as well as metabolic influence and may be due in part to altered sympathetic drive. (Changes in sympathetic drive might be related to the circadian rhythm documented for circulating catecholamines.) It is not reflected in cardiac output which, in a study by Davies and Sargeant [1975], showed no circadian variation. Physiological variables have been examined postexercise, especially for short periods after incremental exercise tests to exhaustion. The elevation in oxygen consumption for 3 min following a $\dot{V}_{O_2}$ max exercise protocol was found to be invariant with time of day [Reilly and Brooks, 1982]. The circadian rhythm in heart rate is evident soon after maximal exercise ceases [Reilly et al., 1984]. Fitness indices such as the Harvard Test Score could therefore have an error as large as 5% due to the time of day the test is performed.

A rhythm in maximal aerobic power cannot therefore be used to explain time of day effects in all-out performance such as 400 m swim trials and 5 min runs. Reilly and Baxter [1983] investigated whether exercise to voluntary exhaustion at an exercise intensity close to $\dot{V}_{O_2}$ max exhibited circadian variation. Subjects performed the task on a cycle ergometer at 06:30 h and at 22:00 h. They were capable of exercising for longer in the evening than in the morning, mean values being 436 and 260 s, respectively. Besides, they tolerated higher blood lactate levels in the evening, the greater levels being a result of the increase in total work performed at that time.

It is tempting to suggest from these results that rhythms in vigorous exercise performance may result from a combination of motivation and psychological drive that are reflected in anaerobic power output. Studies using the Wingate Anaerobic Test [Reilly and Down, 1986b] have failed to show a significant circadian rhythm in anaerobic power and anaerobic capacity for arm or leg exercise. Part of the reason for the absence of rhythmicity may have been that the vigorous warm-up procedures prior to

testing were effective in swamping underlying rhythms. It is also possible that
the sensitivity of such tests is insufficient to detect rhythms that are small in
magnitude.

This insensitivity may extend to computer-controlled devices for measuring maximal muscle function. Isometric muscle strength is known to vary
with time of day in close accordance with the curve in deep body temperature. This applies to grip strength and to back strength. The study of Cabri et
al. [1988] did not find a significant circadian rhythm in peak torque for slow
and fast isokinetic movements under concentric and eccentric conditions.
Nor was there a significant time of day result for a 60-second fatigue test, the
ratio of mean torque for the first 5 contractions and mean torque over the last
5 contractions constituting the 'fatigue index'. It was noted that diastolic
pressure was consistently more perturbed at night than during the daytime. It
was thought that this could have repercussions for safety of exercise performed by coronary risk individuals in the early morning. This 'shock theory'
of exercise – presented as a brief bout of intense effort – would place unfit
individuals, with coronary heart disease predispositions, at greater risk when
exercising abruptly in the morning compared to later on in the day. This view
is supported by findings of Yasue et al. [1979] of a circadian variation in
exercise capacity of patients with variant angina caused by coronary arterial
spasm: this was induced by exercise in the early morning but not in the
afternoon.

Subjective Strain

A circadian variation in the subjective reaction to exercise might explain
why performance is generally better in the evening than in the morning. The
perception of effort does not seem to vary at light ergometric loads up to 150
W, though the slope of the relationship between heart rate and perceived
exertion does change with time of day [Reilly et al., 1984]. Faria and
Drummond [1982] obtained ratings of exertion at treadmill running speeds
eliciting heart rates of 130, 150 and 170 beats min^{-1}. As the heart rate
response to a fixed submaximal exercise intensity is lowest at night, it follows
that more exercise can be performed at a given heart rate at that time. The
higher subjective ratings reported at night may have been due to the work-
rate and not an inherent variation in effort perception.

It is possible that a circadian variation in the rating of exertion is evident
only at high levels of exercise. This is supported by the findings of Reilly and

Baxter [1983] whose subjects first cycled at 40% $\dot{V}_{O_2}$ for 5 min before cycling to exhaustion at 95% $\dot{V}_{O_2}$ max at two different times of day, 06:30 h and 22:00 h. No difference in perceived exertion with time of day was noted at either work-rate, although the higher work-rate was sustained for longer in the evening to the point where the same end-point (exhaustion) was reached. The findings were interpreted as offering support for the concept of a motivational component to the circadian variation in exercise performance.

Reilly and Hales [1988] studied the effects of partial sleep loss on perceived exertion at different times of day. Submaximal exercise was perceived to be harder in the morning than in the evening under those conditions. The time of day effect was greater in magnitude than was the effect of sleep loss. The results suggest that changes in the subjective reactions to exercise with the time of day may be tied more closely to the rhythm in arousal than to that of body temperature.

A time of day effect on the perception of effort was also noted by Wilby et al. [1987]. A standard circuit of weight training was rated as harder when conducted at 07:30 h compared to 22:00 h. In this instance the perceived exertion was related to back strength ($r = -0.59$), back strength being higher in the evening than in the morning. The fact that a set exercise regimen is perceived as harder in the morning compared to the evening may have implications for timing of training. It would be logical to suggest that greater training loads would be tolerated in the evening compared to earlier in the day. Caution should be expressed before arriving at such generalizations, since effects of warm-up, occupational activity during the day, individual characteristics and the nature of the training stimulus need to be considered.

An alternative paradigm which employs light work-bouts every 4 h for 24 h may include a cumulative fatigue effect. With this experimental protocol exercise was rated harder at night-time, and perceived exertion was marginally elevated in the afternoon [Reilly and Young, 1982]. This may represent a sub-harmonic in the circadian rhythm of biological arousal, which is sometimes reflected in a post-lunch dip in performance.

There is also a circadian variation in pain perception that might be relevant in the context of sport injury. Minimum levels of pain are noted in the morning, with a more or less steady increase during the day to a peak during the evening. Laboratory investigations have shown similar findings, subjects becoming more sensitive and the pain perception threshold falling as the day proceeds. The highest threshold for epicritic pain was noted at about 03:00–06:00 h, that for more diffuse pain at 11:00 h [Baxter, 1987].

These observations have implications for the prescription of analgesics for a variety of sports injuries and other painful conditions.

Thermoregulation

The rise in body temperature during exercise may limit endurance performance, especially when the exercise is conducted in the heat. During strenuous exercise the cardiac output has a thermoregulatory function, distributing blood to the skin for cooling. This may compromise blood flow to the active muscles but will not necessarily do so if cardiac output is submaximal. There is no convincing reason to suppose that the highest core temperature that can be tolerated without heat injury does vary with time of day. The guidelines for avoiding heat stress operate implicitly or explicitly on the basis of absolute core temperature values for the stages of heat illness. This would mean that the margin from resting baseline to the risk threshold is about 0.5 °C greater in the morning than in the afternoon. This supports the argument for starting marathon races in the morning rather than in the afternoon in hot climates, which is based on the lower environmental heat stress in the morning. The ideal ambient temperature for marathon running is about 13 °C [Maughan, 1990].

One advantage of starting sustained exercise at a sub-optimal core temperature is that the onset of heat stress is retarded and the overall strain on thermoregulatory mechanisms is reduced. Support for this view was provided by the results of Hessemer et al. [1984]. Pre-cooling esophageal temperature by 0.4 °C and mean skin temperature by 4.5 °C before 60 min of submaximal exercise caused a 6.8% overall increase in work-rate compared with control conditions. An experiment is currently underway to test the significance of the lower body temperature in the morning for sustained exercise performance, the hypothesis being that fatigue due to thermoregulatory factors is delayed if exercise is conducted in the morning.

Our investigations have shown closely related cyclical changes in rectal and skin temperatures. The time course of skin temperature changes varied with the body surface location and the rhythm in the exercising limb tended to disappear during exercise [Reilly and Brooks, 1986]. The observations must be interpreted in the context of an interplay between heat loss and heat gain mechanisms. The data suggest that it is the regulation of body temperature rather than the loss or gain of heat that varies with time of day.

Physical Factors

Physical factors known to vary with time of day and affect muscular function are joint stiffness and flexibility. The former refers to the internal resistance to movement while the latter indicates range of movement about a joint. Stiffness is greatest late in the evening and early in the morning [Wright et al., 1969]. It is affected both by the pattern of activity during the day and the diurnal curve in joint temperature.

Peak values in flexibility have been noted around early afternoon (13:30 h) and trough values first thing in the morning. Specific flexibility factors vary in acrophase; the mean of a number of flexibility measures peaked at 18:10 h in the studies of Gifford [1987]. Findings suggest that abrupt exercise needs to be preceded by warm-up particularly in the morning.

The weight-bearing associated with activity during the day does have consequences as the day progresses. One is the shrinkage in spinal length which from morning to night-time can approach 19 mm in males and 15 mm in females [Reilly et al., 1984; Wilby et al., 1987]. The intervertebral discs stiffen as they lose height, rendering them more vulnerable to injury. Changes in stature that reflect spinal shrinkage may be measured with a sensitive apparatus which controls the posture and spinal contours of subjects [Troup et al., 1985].

Wilby et al. [1987] compared the losses of height as a result of a 20-min circuit weight-training regimen at 07:30 and 23:00 h. The disc was a more effective shock absorber in the morning when shrinkage was 5.4 mm compared to a mean value of 4.3 mm in the evening. This greater stiffness in the evening is partly compensated for by greater back muscle strength. A high negative correlation was found between back strength and height lost, the greater the muscle strength the less height that was lost.

The habitual activity level and the postures engaged in can influence the amount of shrinkage during the day. Intervention procedures for unloading the spine prior to heavy physical training in the evening have been advocated. An example is the Fowler position at rest with the trunk supine and the legs raised to rest on a bench. Gravity inversion systems have also proved effective, although the spinal distension induced pre-exercise is quickly lost once a strenuous exercise regimen is undertaken [Boocock et al., 1988].

Desynchronization

Rhythms are externally desynchronized if the individual switches to a nocturnal work schedule or rapidly crosses a series of time zones. Physiological rhythms affected include body temperature, the sleep-wake cycle, pulse rate, ventilation, arterial pressure, diuresis and excretion of electrolytes [Winget et al., 1985]. The disorientation that results can impair performances of athletes on busy competitive schedules until the major rhythms have all adjusted to the new time zone [De Looy et al., 1988].

Westward travel tends to be easier to adjust to than is traveling in an easterly direction. This probably reflects the fact that in 'free-wheeling' conditions where the major environmental time-givers are absent, the period is lengthened to 25–27 h [Minors and Waterhouse, 1981a]. Performance of psychomotor and athletic skills tends to suffer more following eastward than after westward flights. The differences may be negligible if the time-zone shift is near maximal (fig. 4). This is illustrated in the rate of adaptation of core temperature during the course of an outward and return journey from England to Australia [Reilly, 1990].

Data on rugby players after traveling to Australia clearly show a deterioration in muscular strength until jet-lag symptoms abated. The circadian rhythm reappears once the body has adjusted to local time. Our recommendations are that training is best performed in the morning (having traveled from Europe to Australia) until the body's rhythms are resynchronized [Reilly and Mellor, 1988].

Altering bedtime for a few days before departure and for a few days after arrival may lessen the disruptive effects of traveling across time zones. The changes in sleep schedules should correspond with the direction of intended travel, so that the change of circadian rhythms is smooth. Since only the behavior and sleep-wake cycles are adjusted, the alterations in phase of the new biological rhythms occur at a slower rate than in time-zone transition [Reilly and Maskell, 1989]. Besides, performance is likely to be depressed during the period in which phase shifts are being deployed prior to traveling abroad. Sleep, rather than meal times or social activity, is the main synchronizer of the rhythms: prolonged naps at the new location should be avoided as these would operate against adaptation by anchoring the rhythm at its previous phase [Minors and Waterhouse, 1981b]. Anchoring circadian rhythms is a tactic used by non-sports personnel regularly crossing time zones for short spells abroad: its use by athletes would be practical when they fly in for single contests but this might be counteracted by the effects of travel fatigue.

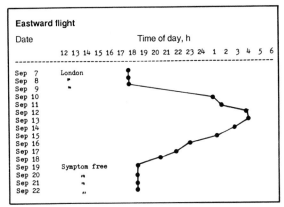
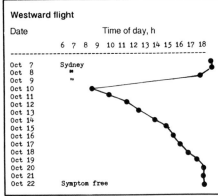

Fig. 4. The time of day at which peak oral temperature was noted is shown according to local time in eastward (out) and westward (return) journey between England and Australia. A stopover in Singapore for 1 night was taken during both journeys. The points were obtained by means of cosinor analysis of data collected throughout each day.

Sleep-Wake Cycles and Sleep Loss

The sleep-wake cycle in humans is reflected in the habits of the majority of the population who sleep at night and work in the daylight hours. Sleep consists of a cycle of stages that recurs about every 90 min. There is not a consensus among sleep researchers about the essential function of sleep [Reilly, 1986]. Leading theories of sleep are related either to tissue restitution or avoidance of exhaustion. Tissue restitution theories [Oswald, 1980] have been linked with heightened mitosis during sleep and elevated growth hormone secretion during slow wave sleep (a combination of stages 3 and 4 of what is described as non-REM sleep, where the rapid eye movements associated with dreaming are absent). As tissue restitution proceeds while awake, even after exhausting exercise, and convincing evidence such as decreased protein turnover in sleep loss is lacking, it is also argued that the restitution processes of sleep may be specific to the needs of nerve cells [Horne, 1988].

There is an interaction between sleep loss and circadian rhythms in that impairments in human performance during sleep deprivation are most pronounced at night-time. In self-paced activity comprised of 4-a-side indoor soccer sustained for 3–4 days, the activity level was found to peak at about 18:00 h, coinciding with the daily high point of body temperature [Reilly and

Walsh, 1981]. Other variables that followed this curve included grip strength and choice reaction time. The circadian rhythm persists for as many days as the individual can be kept awake. Over 3–5 days of total sleep deprivation, there is a trend towards deteriorating performance upon which the circadian rhythm is superimposed [Thomas and Reilly, 1976]. This trend has to be statistically removed prior to establishing circadian rhythms. The trend is not evident in all functions: gross muscular performances such as isometric strength, are highly resistant to effects of sleep loss, despite the decrease in muscle enzyme activity noted after the first night [Vondra et al., 1981]. In contrast, cognitive functions are easily affected. Complex and challenging tasks are less affected than monotonous repetitive ones and strong motivation may overcome the effects of sleep loss, at least for short periods [Wilkinson, 1969].

Subjects deprived of sleep for 2–3 nights begin to exhibit psychotic-like symptoms and bizarre behavior. In such circumstances meaningful physical activity becomes difficult to sustain without error. It has been suggested that naturally occurring brain amines may play a role in the cycles of behavior and mood associated with prolonged sleep loss. Reilly and George [1983] found a circadian rhythm in phenylethylamine levels in urine of sleep-deprived indoor footballers: by the third successive night without sleep, the concentrations of this substance being excreted were found to approach the values typically observed in psychiatric patients. Such prolonged periods of sleeplessness are fortunately only rarely experienced. In military recruits where such regimens are imposed during training, similar trends superimposed on a circadian rhythm in catecholamine excretion have been noted. The curve in catecholamines coincided with the rises and falls in performance of shooting [Akerstedt, 1979].

Partial sleep loss or disrupted sleep is a more common problem. In view of the variability between individuals in the usual amount of sleep taken, tolerance of sleep loss, sensitivity of laboratory measures of performance to effects of sleep deprivation and so on, inferences from experimental investigations must be made with caution. Effects of partial sleep loss also depend on motivation, task complexity, stage of sleep most affected and other factors.

Faculties associated with slow wave sleep may be unimpaired unless the duration of sleep is 3 h or less. Consequently, Reilly and Deykin [1986] examined the effects of a nightly ration of 2.5 h sleep on a battery of psychomotor, work capacity and mental state tests over 3 nights of sleep loss and after 1 night of subsequent recovery. A 3-day control period was used in a counterbalanced design to eliminate an order effect. Functions that required fast reactions were found to deteriorate significantly. This applied to anaerobic power output in a stair run, choice reaction time at rest and during exercise on an ergometer. Exercise attenuated the effects of sleep loss on reaction time, suggesting the benefits of manipulating arousal level by means of a warm-up. Limb speed, as measured by a reciprocal tapping task, got steadily worse over the days of the experiment. Gross motor tasks such as grip strength, lung function and treadmill run time proved resistant to effects of the restricted sleep ration. The single night of recovery sleep was sufficient to reverse the impairments in performance that were observed. As the restricted sleep regimen was found to affect the more complex motor co-ordination tasks while leaving gross motor functions relatively intact, the data were seen to support the 'nerve restitution' theory of sleep.

The effects of partial sleep loss found in males have been replicated in female subjects by Reilly and Hales [1988]. Their subjects were also limited to 2.5 h sleep for 3 successive nights in a counterbalanced experimental design: performances were measured each morning (0-7:00–09:00 h) and each evening (19:00–21:00 h). The time of day significantly influenced the majority of measures; for gross motor function this effect was greater than that of sleep loss. The perceived exertion during cycling at 60% $\dot{V}_{O_2}$ max showed both a diurnal rhythm and an underlying trend towards increased subjective strain. This coincided with the observations on the sleepiness of subjects, self-rated pre-exercise. The reduced subjective ratings on the final experimental morning may be explained by anticipation of the end of the experiment.

That sleep is needed more for 'brain restitution' rather than for 'tissue restitution' was further supported in a study of effects of partial sleep deprivation on swimmers [Sinnerton and Reilly, 1990]. The sleep ration was restricted to 2.5 h a night for 3 consecutive nights. Performance over 400 m and over four successive 50-meter swims was maintained throughout the experimental period. Swimming times were faster in the evening (17:30 h) compared to morning (06:30 h), replicating the findings of Reilly and Hales [1988] that the time-of-day effect on gross motor functions exceeds that of sleep loss. The most pronounced effects of the restricted sleep regimen were deteriorations in mood as the investigation progressed.

A paradoxical finding in studies of partial sleep loss is that some tasks show an improvement. Hand steadiness, for example, is generally better after loss of sleep [Reilly and Deykin, 1986]. This has been attributed to a decrease in spontaneous contraction of involved muscles, due to reduced muscle tone.

Similarly, tasks with high loadings on short-term memory appear to improve with sleep deprivation, due to a tendency to code information acoustically for mental storage and recall in laboratory tests [Reilly et al., 1982]. Thus, care is needed in choosing dependent variables in sleep deprivation studies.

Conclusion

This review has highlighted many biological functions which affect muscle activity according to the time of day. In the main, motor performance rhythms are close in phase to the curve in body temperature. The rhythm in arousal is also an important determinant of performance as it affects motivation towards strenuous exercise and the subjective reactions to it.

The influence of circadian rhythms on performance is most apparent when they are desynchronized during jet lag. They are also evident during sleep loss which gross muscle functions can tolerate relatively well. It seems that motor coordination tasks are likely to break down during sleep loss due to the need for nerve cell restitution. The importance of these cycles should be recognized by sports practitioners.

References

Akerstedt, T.: Altered sleep/wake patterns and circadian rhythms. Acta physiol. scand. suppl. 469 (1979).

Åstrand, P.O.; Rodahl, K.: Textbook of work physiology (McGraw-Hill, New York 1986).

Baxter, C.: Low back pain and time of day: A study of their effects on psychophysical performance; unpubl. doctoral thesis, University of Liverpool (1987).

Baxter, C.; Reilly, T.: Influence of time of day on all-out swimming. Br. J. Sports Med. *17:* 122–127 (1983).

Boocock, M.G.; Garbutt, G.; Reilly, T.; Linge, K.; Troup, J.D.G.: Effect of gravity inversion on exercise induced spinal loading. Ergonomics *31:* 1631–1638 (1988).

Cable, T.; Reilly, T.: Influence of circadian rhythms on arm exercise. J. hum. Mov. Stud. *13:*13–27 (1987).

Cabri, J.; Clarys, J.P.; De Witte, B.; Reilly, T.; Strass, D.: Circadian variation in blood pressure responses to muscular exercise. Ergonomics *31:* 1559–1566 (1988).

Colquhoun, W.P.: Biological rhythms and human performance (Academic Press, New York 1971).

Conlee, R.K.; Rennie, M.J.; Winder, W.W.: Skeletal muscle glycogen content: Diurnal variation and effects of fasting. Am. J. Physiol. *231:* 614–618 (1976).

Conroy, R.T.W.L.; O'Brien, M.: Diurnal variation in athletic performance. J. Physiol. *236:* 51P (1974).

Davies, C.T.M.; Sargeant, A.J.: Circadian variation in physiological responses to exercise on a stationary bicycle ergometer. Br. J. ind. Med. *32:* 110–114 (1975).

De Looy, A.; Minors, D.; Waterhouse, J.; Reilly, T.; Tunstall-Pedoe, D.: The coach's guide to competing abroad. (National Coaching Foundation, Leeds 1988).

Faria, I.E.; Drummond, B.J.: Circadian changes in resting heart rate and body temperature, maximal oxygen consumption and perceived exertion. Ergonomics *25:* 381–386 (1982).

Folkard, S.: Diurnal variation in logical reasoning. Br. J. Psychol. *66:* 18 (1975).

Gifford, L.S.: Circadian variation in human flexibility and grip strength. Aust. J. Physiother. *33:* 3–9 (1987).

Hessemer, V.; Langusch, D.; Bruck, K.; Bodeker, R.K.; Breidenbach, T.: Effects of slightly lowered body temperature on endurance performance in humans. J. appl. Physiol. *57:* 1731–1737 (1984).

Horne, J.A.: Why we sleep: The function of sleep in humans and other mammals (Oxford University Press, Oxford 1988).

Maughan, R.J.: The marathon; in Reilly, Secher, Snell, Williams, Physiology of sports, pp. 121–152 (Spon, London 1990).

Minors, D.S.; Waterhouse, J.M.: Circadian rhythms and the human (John Wright, Bristol 1981a).

Minors, D.S.; Waterhouse, J.M.: Anchor sleep as a synchroniser of abnormal routines. Int. J. Chronobiol. *7:* 165–188 (1981b).

Oswald, I.: Sleep as a restorative process: Human clues. Progr. Brain Res. *53:* 279–288 (1980).

Reilly, T.: Circadian variation in ventilatory and metabolic adaptations to submaximal exercise. Br. J. Sports Med. *16:* 115–116 (1982).

Reilly, T.: Exercise and sleep: An overview; in Watkins, Reilly, Burwitz, Sports science, pp. 414–419 (Spon, London 1986).

Reilly, T.: Circadian rhythms and exercise; in Macleod, Maughan, Nimmo, Reilly, Williams, Exercise: benefits, limits and adaptations, pp. 346–366 (Spon, London 1987).

Reilly, T.: Human circadian rhythms and exercise. Clin. Rev. Biomed. Eng. *18:* 165–180 (1990a).

Reilly, T.: Time zone shift and sleep deprivation problems; in Torg, Welsh, Shephard, Current theory in sports medicine – 2, pp. 135–139 (Decker, Toronto 1990b).

Reilly, T.; Baxter, C.: Influence of time of day on reactions to cycling at a fixed high intensity. Br. J. Sports Med. *17:* 128–130 (1983).

Reilly, T.; Brooks, G.A.: Investigation of circadian rhythms in metabolic responses to exercise. Ergonomics *25:* 1093–1197 (1982).

Reilly, T.; Brooks, G.A.: Exercise and the circadian variation in body temperature measures. Int. J. Sports Med. *7:* 358–362 (1986).

Reilly, T.; Brooks, G.A.: Selective persistence of circadian rhythms in physiological responses to exercise. Chronobiol. Int. *7:* 59–67 (1990).

Reilly, T.; Deykin, T.: Effects of partial sleep loss on subjective states, psychomotor and physical performance tests. J. human Mov. Stud. *9:* 157–170 (1986).

Reilly, T.; Deykin, T.; Wilkinson, R.T.: Influence of sustained exercise without sleep on state anxiety and reaction time. Proc. 20th Int. Congr. Appl. Psychology (Clinical and Physiological Psychology), Edinburgh, 1982, p. 410.

Reilly, T.; Down, A.: Circadian variation in the standing broad jump. Percept. Mot. Skills *62:* 830 (1986a).

Reilly, T.; Down, A.: Time of day and performance on all-out arm ergometry; in Reilly, Watkins, Borms, Kinanthropometry III, pp. 296–300 (Spon, London 1986b).

Reilly, T.; George, A.: Urinary phenylethylamine levels during three days of indoor soccer play. J. Sports Sci. *1:* 70 (1983).

Reilly, T.; Hales, A.J..: Effects of partial sleep deprivation on performance measures in females; in Megaw, Contemporary Ergonomics 1988, pp. 509–514 (Taylor Francis, London 1988).

Reilly, T.; Maskell, P.: Effects of altering the sleep-wake cycle in human circadian rhythms and motor performance. Proc. 1st IOC Congr. Sport Sci., Colorado Springs, 1989, pp. 106–107.

Reilly, T.; Mellor, S.: Jet lag in student Rugby League players following a near-maximal time-zone shift; in Reilly, Lees, Davids, Murphy, Science and football, pp. 249–256 (Spon, London 1988).

Reilly, T.; Robinson, G.; Minors, D.S.: Some circulatory responses to exercise at different times of day. Med. Sci. Sports Exerc. *16:* 477–482 (1984).

Reilly, T.; Tyrrell, A.; Troup, J.D.G.: Circadian variation in human stature. Chronobiol. int. *1:* 121–126 (1984).

Reilly, T.; Walsh, T.J.: Physiological, psychological and performance measures during an endurance record for 5-a-side soccer play. Br. J. Sports Med. *15:* 122–128 (1982).

Reilly, T.; Young, K.: Digit summation, perceived exertion and time of day under submaximal exercise conditions. Proc. 20th Int. Congr. Appl. Psychology, Edinburgh, 1982, p. 344.

Rodahl, A.; O'Brien, M.; Firth, P.G.R.: Diurnal variation in performance of competitive swimmers. J. Sports Med. Phys. Fit. *16:* 72–76 (1976).

Shephard, R.J.: Sleep, biorhythms and human performance. Sports Med. *1:* 11–37 (1984).

Sinnerton, S.; Reilly, T.: Effects of sleep loss and time of day in swimming. Commun. VIth Int. Symp. Biomechanics and Medicine in Swimming, Liverpool, 1990.

Stockton, I.D.; Reilly, T.; Sanderson, F.H.; Walsh, T.J.: Investigations of circadian rhythms in selected components of sports performance. Commun. Soc. Sports Sciences, Crewe and Alsager College, 1978.

Thomas, V.; Reilly, T.: Circulatory, psychological and performance variables during 100 hours of paced continuous exercise under conditions of controlled energy intake and work output. J. hum. Mov. Stud. *1:* 149–155 (1975).

Troup, J.D.G.; Reilly, T.; Eklund, J.A.E.; Leatt, P.: Changes in stature with spinal loading and their relation to the perception of exertion or discomfort. Stress Med. *1:* 303–307 (1985).

Vondra, K.; Brodan, V.; Bass, A.; Kuhn, E.; Tersinger, J.; Andel, M.; Veselkova, A.: Effects of sleep deprivation on the activity of selected metabolic enzymes in skeletal muscle. Eur. J. appl. Physiol. *47:* 41–46 (1981).

Wilby, J.; Linge, K.; Reilly, T.; Troup, J.D.G.: Spinal shrinkage in females: Circadian variation and the effects of circuit weight-training. Ergonomics *30:* 47–54 (1987).

Wilkinson, R.T.: Some factors influencing the effect of environmental stressors upon performance. Biol. Bull. *72:* 260–272 (1969).

Winget, C.M.; De Roshia, C.W.; Holley, D.C.: Circadian rhythms and athletic performance. Med. Sci. Sports Exerc. *17:* 498–516 (1985).

Wit, A.: Zayanienia regulacji w procesie rozwoju sily miesnionej na przykladzie zawodnikow uprawiajacych podnoszenie ciezarow (Institute of Sport, Warsaw 1980).

Wright, V.; Dawson, D.; Longfield, M.D.: Joint stiffness – its characterisation and significance. Biol. med. Eng. *4:* 8–14 (1969).

Yasue, H.; Omote, S.; Takizawa, A.; Nagao, M.; Miwa, K.; Tanaka, S.: Circadian variation of exercise capacity in patients with Prinzmetal's variant angina: Role of exercise-induced coronary arterial spasm. Circulation *59:* 938–948 (1979).

Zimmet, P.Z.; Wall, J.R.; Rome, L.; Stimmler, L.; Jarrett, R.J.: Diurnal variation in glucose tolerance: associated changes in plasma insulin, growth hormone and non-esterified fatty acid. Br. med. J. *ii:* 485–488 (1974).

Thomas Reilly, Centre for Sport and Exercise Sciences,
Liverpool Polytechnic, Byrom Street, GB–Liverpool L3 3AF (UK)

Subject Index